Preface

This Study Guide is an ancillary text to Basic Engineering Circuit Analysis, 8th Edition, hereafter referred to as BECA, by J. David Irwin and R. Mark Nelms. This Study Guide treats computer simulation techniques in more depth than could be covered in the text. The computer simulation techniques included are for PSpice®, MATLAB®, and Microsoft ® Excel. In addition to the simulation techniques, the Study Guide also includes some special sections that are designed to enhance the reader's understanding of some practical applications, e.g. the display of a typical power company's distribution system in Section 14.

The CD-ROM inside the front cover of this Study Guide includes most of the abovementioned simulations and demonstrations. Also on the CD-ROM is a folder called Visual Tutors. The files in this folder are executable videos that demonstrate some of the key PSpice examples within this Study Guide. Since Visual Tutor files are executable, you don't need video software to view them. Simply run them as you would any other executable file. Finally, the CD contains a folder called BECA libraries which contains four PSpice library files. These libraries contain some custom parts we have created for your enjoyment. Procedures for adding the BECA libraries to PSpice are discussed in Section of the guide. Each of the major file types on the CD and the software version used to create them are listed below. Compatibility with older or newer versions is not guaranteed.

PSPICE version 9.1
Microsoft Excel version 7.0
MATLAB version 5.1

Trademarks

PSpice is a registered trademark of Cadence Design Systems, Inc.
Microsoft® Excel is a registered trademark of Microsoft Corp.
MATLAB is a registered trademark of The Math Works, Inc.

Table of Contents

(**BECA Chapter** refers to the text *Basic Engineering Circuit Analysis, 8th Edition*, indicating with which text chapters the chapters of the Study Guide are related.

BECA Chapter 14 – Application of the Laplace Transform to Circuit Analysis

1. The Passive Sign Convention

One of the fundamental laws in circuit analysis, Ohm's Law, states that the voltage-current relationship for a resistor is

$$V = IR \qquad\qquad (1.1)$$

There is however an underlying requirement for Ohm's Law to be used properly, the *passive sign convention*. This convention, demonstrated in Figure 1.1, sets the relationship between the polarity of the voltage, V, and the direction of the current, I. In the passive sign convention, current is ASSUMED to enter the positive side of the voltage polarity. Each of the circuits in Figure 1.1 represent the exact same situation: 2 Amps of current

flowing from node A to node B resulting in a potential of 20 Volts at node A with respect to node B. Applying Ohm's Law, we find the resistance value is 10 Ω in each case.

The passive sign convention is absolutely critical in determining power flow in circuit analyses. We know that the power consumed or generated by an element is

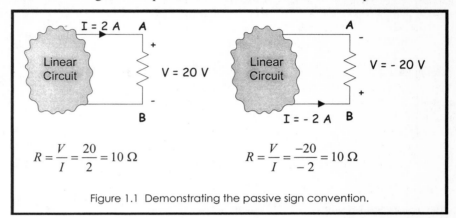

Figure 1.1 Demonstrating the passive sign convention.

$$P = VI \qquad\qquad (1.2)$$

But this equation tells us nothing about the power flow, generation or consumption? As the name implies, when using the passive sign convention definitions for current direction and voltage polarity, a positive result for P means power is being consumed – the element is passive. Conversely, if P is negative, the element is supplying power. Consider the circuit in Figure 1.2 where the passive sign convention has been applied to all three elements. Based on our calculations, the resistor and the 1-V source are using power while the 10-V source delivers power.

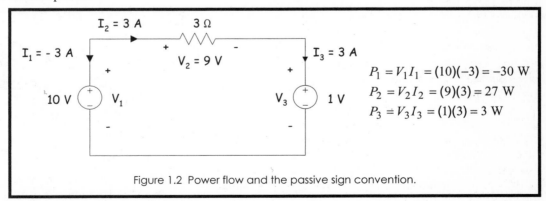

Figure 1.2 Power flow and the passive sign convention.

1

2. Resistor Construction

Resistors are the most basic and numerous circuit elements. Commercial resistors obey Ohm's Law quite well, are available in values from mΩ (1/1000) to GΩ (10^9), cost less than $0.05 each (in 2001) and are sold by the millions each month. Two manufacturing processes account for the majority of the general purpose resistor market – film and wirewound technology.

Figure 2.1 shows a film resistor in various stages of construction. The principle components are the ceramic bobbin, the film, the end caps and the encapsulation. The ceramic bobbin is an insulator, conducting no current. It serves only as a mechanical support for the film layer which is the true resistive material.

Film materials fall into two categories: metal films, sometimes called thin films, and carbon based films sometimes called thick films. In metal film resistors, a poorly conducting metal such as manganin™ or nichrome™ is deposited onto the bobbin in a very thin layer, about 1 millionth of a meter. External wires, called leads, are attached to the film via the metal end caps. Finally, the body of the resistor is coated in a plastic encapsulant for mechanical and environmental protection. In carbon film resistors, the film starts as a slurry containing carbon (insulator) and metal (conductor) particles. The bobbin is coated with the mixture, then it is cured in an oven. End caps and encapsulant are added to finish the process.

How is the resistor's value controlled? First consider the resistive block in Figure 2.2. The resistance of any piece of material can be expressed as

$$R = \frac{\rho L}{tw} = \frac{\rho L}{A} \qquad (2.1)$$

where w is the width, t is the thickness, A is the cross-sectional area, L is the length and ρ is the resistivity of the material. The resistivity of the metal in metal film resistors is determined by which metal is used. Carbon film resistivity is controlled by adjusting the ratio of carbon to metal in the slurry. After deposition, the film can be ablated using a laser or a miniature sandblaster to create the serpentine effect in Figure 2.1. This increases the resistor's value by simultaneously increasing L and decreasing w. Ablation also allows the manufacturer to trim the resistor's value to meet tolerance requirements.

Another class of film resistor is the chip resistor, shown in Figure 2.3. It is manufactured using either thin metal films or thick ruthenium-based films on a ceramic base. Used extensively in printed circuit boards, chip resistors attach directly to the surface of the PC board. This eliminates the need for end cap wires, making the part much smaller, lighter and less expensive.

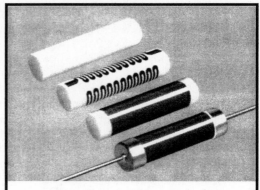

Figure 2.1 A film resistor in various stages of construction.

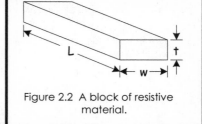

Figure 2.2 A block of resistive material.

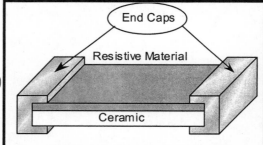

Figure 2.3 Internal diagram of a chip resistor.

Figure 2.4 Internal view of an enamel coated wirewound resistor.

A wirewound resistor, shown in cross-section in Figure 2.4, is little more than a long piece of wire that has been wound up into a coil. To keep the length of the wire reasonable, the wire is made of a poor conducting metal such as nichrome™ ($\rho = 110\ \mu\Omega$-cm). Wirewound resistors, often packaged in silicone or in hollow enamel tubes, are inherently bulkier and more expensive than carbon film resistors. They do however, have two advantages. First, they can be wound to very precise and custom values. Second, they can be physically sized to handle large amounts of current and power. Commercial wirewound resistors range in value from about 0.1 Ω to 1 MΩ.

POWER AND TOLERANCE

Besides the resistance value, there are two other important resistor specifications - tolerance and power rating. The tolerance is the manufacturer's guaranteed range for the resistance. For example, a 1000 Ω, 5% resistor will have an actual resistance between 950 and 1050 Ω. Standard tolerances are 5% and 2%. Tighter tolerances are available but are more expensive.

A resistor's power rating is the maximum continuous power the element can absorb. The problem is that as a resistor absorbs power (energy) – it gets warmer. More power – more heat. Eventually, the heat causes something inside the resistor to give and the part fails. In extreme cases, the resistor catches on fire! Standard power ratings for carbon film resistors are $^1/_8$, ¼, ½ and 1 Watt. Wirewound resistor power ratings are typically between 1 and 200 Watts. As an example of how easy it can be to exceed a power rating, consider an ordinary ¼ W, 5-Ω resistor connected to a fresh 1.5-V, D cell battery. What an innocent scenario. The power absorbed by the resistor is

$$P = \frac{V^2}{R} = \frac{1.5^2}{5} = 0.45 \text{ W} \tag{2.2}$$

almost twice the power rating! So, always consider the expected power dissipation when purchasing resistors.

THE COLOR CODE

Resistor values are often coded using the color code described and demonstrated in Figure 2.5 and Table 2.1. Based on the color code, a 200-kΩ, 5% resistor will have the color code RED-BLACK-YELLOW-GOLD = $20 \times 10^4 \pm 5\%$. This code is used primarily on carbon film resistors and has no information about the part's power rating. Most wirewound resistors have their specifications, resistance, power rating and tolerance, printed directly on the element.

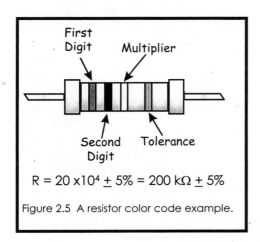

Figure 2.5 A resistor color code example.

R = 20 x10⁴ ± 5% = 200 kΩ ± 5%

TABLE 2.1 THE RESISTOR COLOR CODE

Color	Value	Tolerance
Silver		10%
Gold		5%
Black	0	
Brown	1	
Red	2	2%
Orange	3	
Yellow	4	
Green	5	
Blue	6	
Violet	7	
Gray	8	
White	9	

3. MATLAB Simulations of dc Circuits

3.1 Let us use MATLAB to solve for the current I_O in the circuit in Figure 3.1

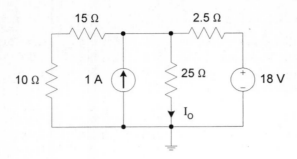

Figure 3.1

Given the circuit has only three nodes (two if we combine the series-connected resistors) and a ground-referenced voltage source, nodal analysis should provide a straightforward solution. The circuit is redrawn in Figure 3.2 where node voltage labels have been applied.

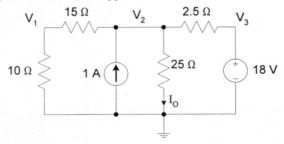

Figure 3.2

At node V_1, KCL yields

$$\frac{V_2 - V_1}{15} = \frac{V_1}{10}$$

or

$$-5V_1 + 2V_2 = 0$$

At node V_2, the KCL equation is

$$1 = \frac{V_2 - V_1}{15} + \frac{V_2}{25} + \frac{V_2 - V_3}{2.5}$$

or

$$-5V_1 + 38V_2 - 30V_3 = 75$$

And, of course, $V_3 = 18$ V. In matrix form, we have

$$\begin{bmatrix} -5 & 2 & 0 \\ -5 & 38 & -30 \\ 0 & 0 & 1 \end{bmatrix} \begin{bmatrix} V_1 \\ V_2 \\ V_3 \end{bmatrix} = \begin{bmatrix} 0 \\ 75 \\ 18 \end{bmatrix}$$

The node voltage vector can be written as

$$\begin{bmatrix} V_1 \\ V_2 \\ V_3 \end{bmatrix} = \begin{bmatrix} -5 & 2 & 0 \\ -5 & 38 & -30 \\ 0 & 0 & 1 \end{bmatrix}^{-1} \begin{bmatrix} 0 \\ 75 \\ 18 \end{bmatrix}$$

In MATLAB, matrices can be input as follows where G is the 3x3 conductance matrix and I is the current vector. We enter the following

```
EDU» I=[0;75;18]
```

and the computer returns

```
I =

     0
    75
    18
```

In a similar manner,

```
EDU» G=[-5 2 0;-5 38 -30;0 0 1]
```

and

```
G =

    -5     2     0
    -5    38   -30
     0     0     1
```

Then the solution is obtained from the equation

```
EDU» V= inv(G)*I
```

which is

```
V =

    6.8333
   17.0833
   18.0000
```

And so, $V_2 = 17.0833$ V and $I_O = V_2/25 = 683.3$ mA.

3.2 Let us use MATLAB to solve for the power delivered by the current source in the circuit in Figure 3.3

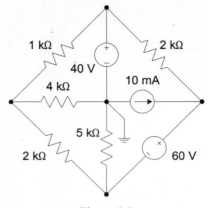

Figure 3.3

Although the circuit has an equal number of nodes and loops, there are two independent voltage sources which produce simple constraint equations in nodal analysis. Thus, we will use nodal analysis to solve this circuit. Given the node voltage definitions and the supernode (denoted by the dashed line) in Figure 3.4, we can write the following nodal expressions.

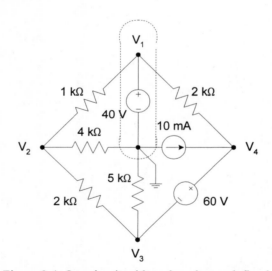

Figure 3.4 Our circuit with node voltages defined.

Node 1: $V_1 = 40$ V   or $V_1 = 40$

Node 2: $\dfrac{V_2 - V_1}{1k} + \dfrac{V_2 - V_3}{2k} + \dfrac{V_2}{4k} = 0$ or $-V_1 + \dfrac{7V_2}{4} - \dfrac{V_3}{2} = 0$

Node 3&4: $V_4 - V_3 = 60$ V   or $-V_3 + V_4 = 60$

Supernode: $\dfrac{V_1 - V_2}{1k} + \dfrac{V_1 - V_4}{2k} + \dfrac{0 - V_2}{4k} + \dfrac{0 - V_3}{5k} = -10m$ or $\dfrac{3V_1}{2} - \dfrac{5V_2}{4} - \dfrac{V_3}{5} - \dfrac{V_4}{2} = -10$

These equations can be written in matrix form as

6

$$\begin{bmatrix} 1 & 0 & 0 & 0 \\ -1 & 7/4 & -1/2 & 0 \\ 0 & 0 & -1 & 1 \\ 3/2 & -5/4 & -1/5 & -1/2 \end{bmatrix} \begin{bmatrix} V_1 \\ V_2 \\ V_3 \\ V_4 \end{bmatrix} = \begin{bmatrix} 40 \\ 0 \\ 60 \\ -10 \end{bmatrix}$$

Solving for the unknowns requires solution of the expression

$$\begin{bmatrix} V_1 \\ V_2 \\ V_3 \\ V_4 \end{bmatrix} = \begin{bmatrix} 1 & 0 & 0 & 0 \\ -1 & 7/4 & -1/2 & 0 \\ 0 & 0 & -1 & 1 \\ 3/2 & -5/4 & -1/5 & -1/2 \end{bmatrix}^{-1} \begin{bmatrix} 40 \\ 0 \\ 60 \\ -10 \end{bmatrix}$$

In MATLAB, we will let G be the 4x4 conductance matrix, I will be the 4x1 vector and the unknown vector will be V. The conductance matrix is placed in MATLAB form as follows

```
» G=[1 0 0 0;-1 7/4 -1/2 0;0 0 -1 1;3/2 -5/4 0 -1/2]
```

and the computer returns

```
G =
      1.0000         0         0         0
     -1.0000    1.7500   -0.5000         0
           0         0   -1.0000    1.0000
      1.5000   -1.2500   -0.2000   -0.5000
```

The I vector is

```
» I=[40;0;60;-10]
```

or

```
I =
     40
      0
     60
    -10
```

The solution is obtained from the equation

```
» V=inv(G)*I
```

which is

```
V =
    40.0000
    25.9459
    10.8108
    70.8108
```

The power delivered by the current source is $P = IV$, where $I = 10$ mA and $V = V_4 = 70.81$ V. Thus, the delivered power is 0.708 W.

3.3 Let us find V_O in the network in Figure 3.5 using nodal analysis and MATLAB.

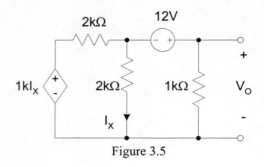

Figure 3.5

The circuit is redrawn in Figure 3.6 with node voltage labels. Since there are three nodes, we need three equations.

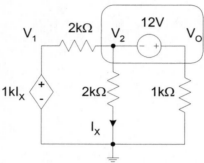

Figure 3.6

The constraint equation for the 12-V source is,

$$V_O - V_2 = 12 \quad \text{or} \quad -V_2 + V_0 = 12$$

The dependent source voltage is controlled by the current $I_X = V_2/2000$, and thus,

$$V_1 = 1000 I_X = \frac{V_2}{2} \quad \text{or} \quad V_1 - \frac{V_2}{2} = 0$$

A third nodal equation is obtained from the supernode shown in Figure 3.6. Summing all current exiting the supernode,

$$\frac{V_2 - V_1}{2000} + \frac{V_2}{2000} + \frac{V_O}{1000} = 0 \quad \text{or} \quad -V_1 + 2V_2 + 2V_O = 0$$

These equations can be written in matrix form as

$$\begin{bmatrix} 0 & -1 & 1 \\ 1 & -0.5 & 0 \\ -1 & 2 & 2 \end{bmatrix} \begin{bmatrix} V_1 \\ V_2 \\ V_O \end{bmatrix} = \begin{bmatrix} 12 \\ 0 \\ 0 \end{bmatrix}$$

Solving for the unknowns requires solution of the expression

$$\begin{bmatrix} V_1 \\ V_2 \\ V_O \end{bmatrix} = \begin{bmatrix} 0 & -1 & 1 \\ 1 & -0.5 & 0 \\ -1 & 2 & 2 \end{bmatrix}^{-1} \begin{bmatrix} 12 \\ 0 \\ 0 \end{bmatrix}$$

In MATLAB, we will let G be the 3x3 conductance matrix, I will be the 3x1 vector and the unknown vector will be V. The conductance matrix is placed in MATLAB form as follows

```
» G=[0 -1 1;1 -0.5 0;-1 2 2];
```

and the computer returns

```
G =
         0   -1.0000    1.0000
    1.0000   -0.5000         0
   -1.0000    2.0000    2.0000
```

The I vector is

```
» I=[12;0;0]
```

or

```
I =
    12
     0
     0
```

The solution is obtained from the equation

```
» V=inv(G)*I
```

which is

```
V =
   -3.4286
   -6.8571
    5.1429
```

Thus, $V_O = 5.1429$ V = 36/7 V.

3.4 Let us use MATLAB and loop analysis to find V_O in the network in Figure 3.7.

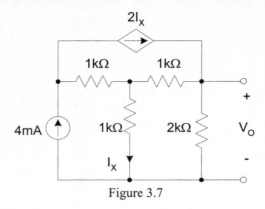

Figure 3.7

The circuit is redrawn in Figure 3.8 with loop currents defined.

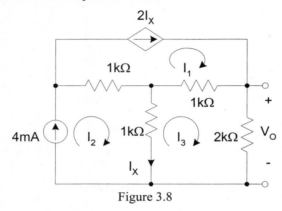

Figure 3.8

For loop 1, we have

$$I_1 = 2I_X \quad \text{or} \quad I_1 - 2I_X = 0$$

where I_X can be written as

$$I_X = I_2 - I_3 \quad \text{or} \quad -I_2 + I_3 + I_X = 0$$

In loop 2, the independent current source forces

$$I_2 = 4\text{mA}$$

Finally, in loop 3,

$$2000I_3 + 1000(I_3 - I_2) + 1000(I_3 - I_1) = 0 \quad \text{or} \quad -I_1 - I_2 + 4I_3 = 0$$

These equations can be expressed in matrix form as

$$
\begin{bmatrix}
1 & 0 & 0 & -2 \\
0 & 1 & 0 & 0 \\
0 & -1 & 1 & 1 \\
-1 & -1 & 4 & 0
\end{bmatrix}
\begin{bmatrix}
I_1 \\
I_2 \\
I_3 \\
I_X
\end{bmatrix}
=
\begin{bmatrix}
0 \\
0.004 \\
0 \\
0
\end{bmatrix}
$$

Solving for the unknowns requires solution of the expression

$$
\begin{bmatrix} I_1 \\ I_2 \\ I_3 \\ I_X \end{bmatrix} = \begin{bmatrix} 1 & 0 & 0 & -2 \\ 0 & 1 & 0 & 0 \\ 0 & -1 & 1 & 1 \\ -1 & -1 & 4 & 0 \end{bmatrix}^{-1} \begin{bmatrix} 0 \\ 0.004 \\ 0 \\ 0 \end{bmatrix}
$$

In MATLAB, we let R be the 4x4 resistance matrix, V is the 4x1 voltage vector and I is the 4x1 current vector. The resistance matrix is placed in MATLAB form as follows

```
» R=[1 0 0 -2;0 1 0 0;0 -1 1 1;-1 -1 4 0]
```

and the computer returns

```
R =

    1     0     0    -2
    0     1     0     0
    0    -1     1     1
   -1    -1     4     0
```

The V vector is

```
» V=[0;0.004;0;0]

V =

        0
   0.0040
        0
        0
```

The solution is obtained from the equation

```
» I=inv(R)*V
```

which is

```
I =

   0.0040
   0.0040
   0.0020
```

From Figure 3.7, $V_O = 2000I_3 = (2000)(0.002) = 4$ V.

4. PSPICE Simulations of dc Circuits

4.1 Let us use PSPICE to determine V_O and I_O in the circuit shown in Figure 4.1.

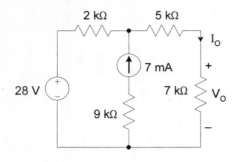

Figure 4.1

From the Start menu, select Start/Programs/PSpice Student/Schematics to open the *Schematics* program. The *Schematics* window, shown partially in Figure 4.2, should appear. Next, we get the necessary parts (dc voltage and current sources and resistors) by selecting Get New Part from the Draw menu. Figure 4.3 shows the Parts Browser window where VDC has been typed into the Part Name field. Selecting Place & Close returns us to the *Schematics* window with the cursor changed to a battery symbol. Left click once to place a voltage source on the page. Right click to return the cursor to its original shape. Your *Schematics* page should look something like Figure 4.4.

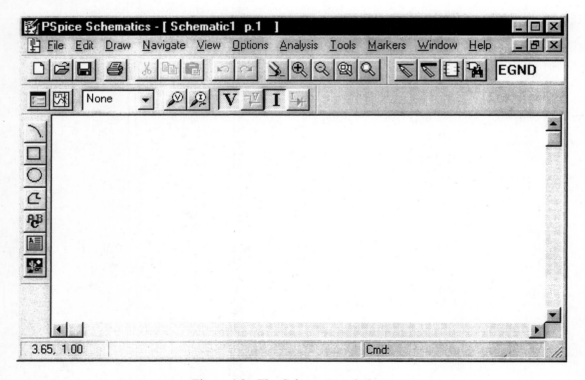

Figure 4.2. The *Schematics* window.

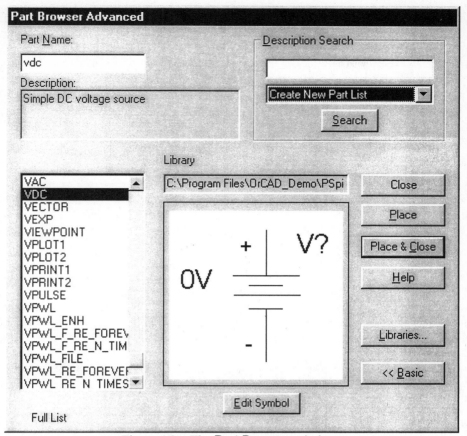

Figure 4.3.   The **Part Browser** window.

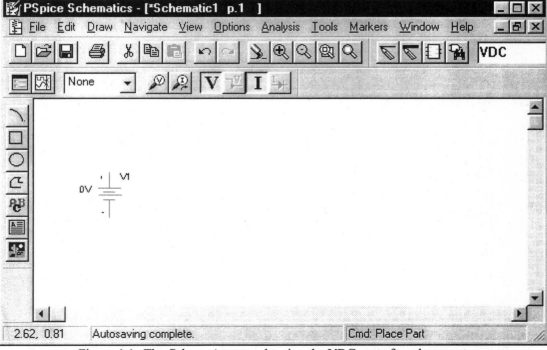

Figure 4.4. The *Schematics* page showing the VDC part after placement.

To get the current source, again select **Get New Part** and type **IDC** into the **Part Name** field as shown in Figure 4.5. This time select **Place** to return to the *Schematics* page and left click to place the current source. The results are shown in Figure 4.6. To get the resistors, move back to the **Parts Browser** window, type **R** in the **Part Name** field and select **Place**. Back in the *Schematics* page, left click to place the first resistor, then move the cursor some distance and left click again to place the second resistor, and so on. Moving the cursor between part placements is important since parts placed atop one another are indistinguishable. Finally, all PSPICE circuits must have a ground terminal. To get one, return to the **Parts Browser** window, type **EGND** in the **Part Name** field and select **Place and Close**. Place the ground part below the other parts. After a right click, the mouse returns to its standard form and the *Schematics* page should look a lot like Figure 4.7.

Now we must orient and move the parts to reasonable locations. To move a part, left click on it once to "select" it. This should change the part's color to red. Next, click and hold on the part, drag it to the desired location and release the mouse button. To rotate a part, select it and choose **Rotate** from the **Edit** menu. This will spin the part 90° counterclockwise. For example, the current source and resistors R2 and R4 in Figure 4.7 should be rotated. From Figure 4.8, we see that after some dragging and rotating, the circuit begins to take shape.

Next, we add wiring to the circuit. From the **Draw** menu, select **Wire**, which changes the cursor to a pencil icon. Let's draw a wire from the top of voltage source V1 to resistor R1. Point the cursor at the end of the connection at the top of V1 and left click once to connect the wire. Move the cursor up and to the right until it reaches the connection at R1. At that point, left click again to connect and cut the wire. The resulting wire is shown in Figure 4.9. Repeat this process to add wires between the remaining parts. Your circuit should look similar to that in Figure 4.10. Take your time when wiring. This is where most students drop the ball.

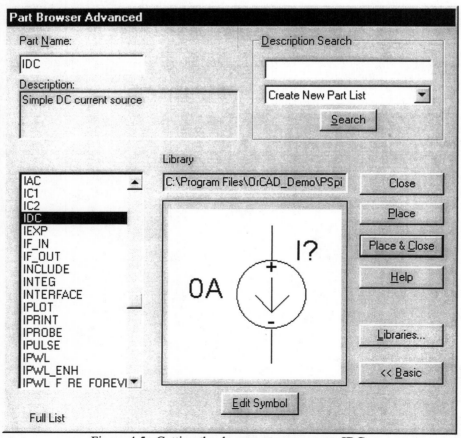

Figure 4.5. Getting the dc current source part, IDC.

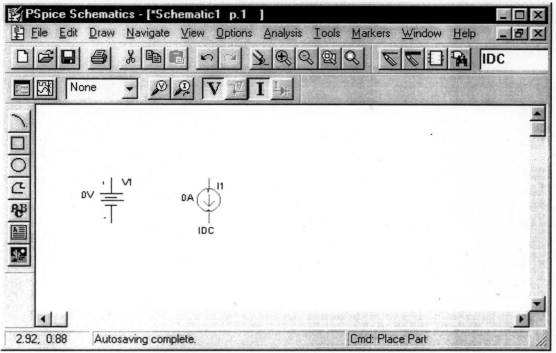

Figure 4.6. The *Schematics* page after placing the IDC part.

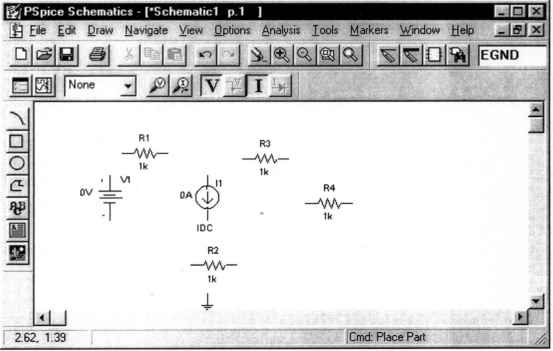

Figure 4.7. The *Schematics* page after all parts have been placed.

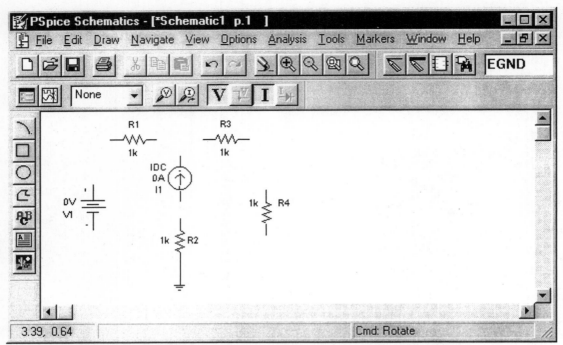

Figure 4.8. The *Schematics* page shows the circuit ready for wiring.

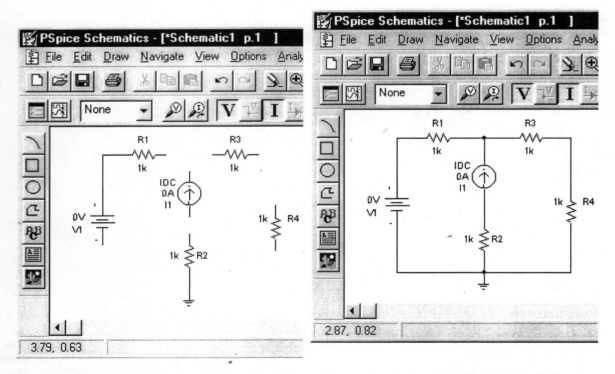

Figure 4.9. The circuit after wiring V1 to R1.   Figure 4.10. The circuit with all wiring finished.

To change the value of R1, double click on its present value (1k) to open the **Attribute** window in Figure 4.11, type in the desired value of 2k and select **OK**. Repeat for all other part values including the voltage and current source. Similarly, to change the label of the voltage source from V1 to Vs, double click on the text "V1" to open the **Reference Designator** window, shown in Figure 4.12, type in the desired name and select **OK**.

16

(Reference Designator is PSPICE-ese for a part's label.) Your circuit, shown in Figure 4.13, is now complete and ready for simulation.

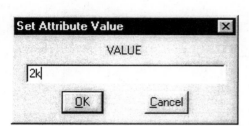

Figure 4.11. The Attribute window is used to change a part's value.

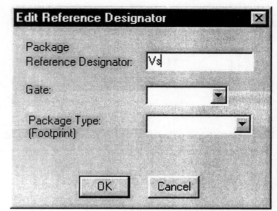

Figure 4.12. The Reference Designator window is used to change a part's label.

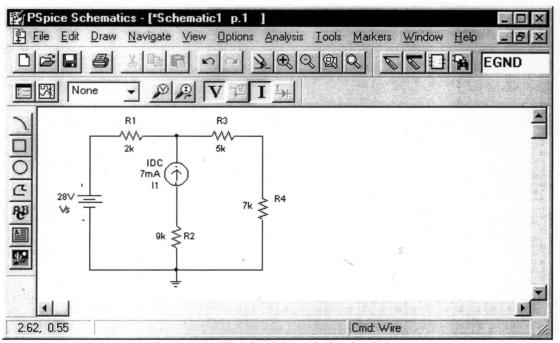

Figure 4.13. The circuit is ready for simulation.

All dc node voltages and currents can be displayed directly on the *Schematics* page. In the Analysis menu, select Display Results on Schematic/Enable Voltage and repeat for Enable Current. Before simulating, select Save in the File menu and save your work. To perform the actual simulation, choose Simulate from the Analysis menu. The PROBE utility will open, displaying the window in Figure 4.14. When the message Simulation Complete appears in the window in the lower left, close PROBE to return to *Schematics*. The results, shown in Figure 4.15, indicate that $V_O = 21$ V and $I_O = 3$ mA. Unwanted voltages and currents can be removed by clicking on them and pressing the Delete key.

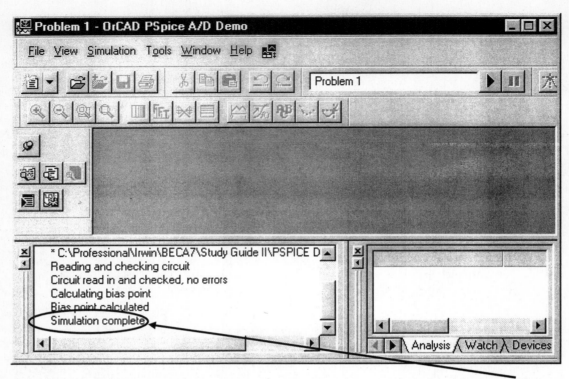

Figure 4.14. When simulating, the PROBE utility opens and informs you when the simulation is finished.

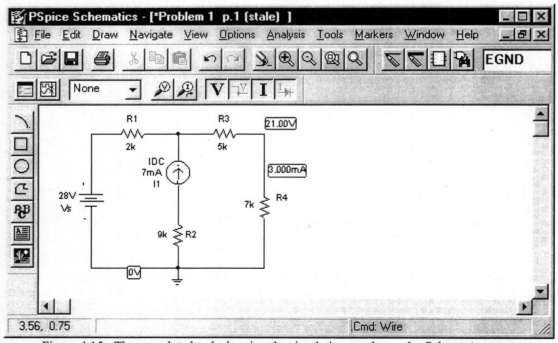

Figure 4.15. The completed task showing the simulation results on the *Schematics* page.

4.2 Let us use PSPICE to determine the current I_{DC} required to force V_O to zero in the circuit in Figure 4.16.

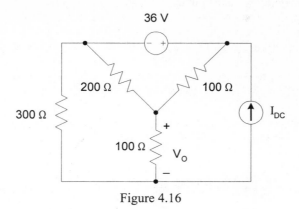

Figure 4.16

Drawing a circuit has been covered step-by-step in both Problem 4.1 and in the main textbook. When you have finished, your *Schematics* page should look something like Figure 4.17, where the current source name has been changed from I_1 to I_{dc}. Note that a voltage marker has been added at the output voltage node. This will inform PSPICE that we wish to view V_O in any plots that might be created.

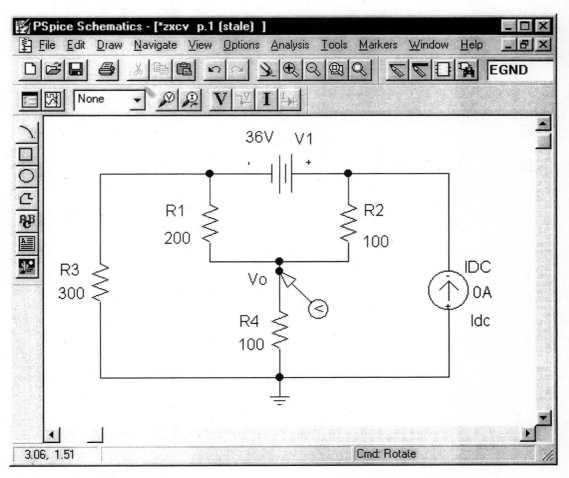

Figure 4.17. The *Schematics* page for the circuit in Problem 4.2.

A very simple way to find the required value of I_{DC} such that V_O equals zero is to sweep I_{DC} across a range of values, plot V_O and, from the zero crossing, extract the I_{dc} value from the plot. To do this, we must request a DC Sweep. Activate the Setup window (Setup in the Analysis menu) and select DC Sweep, as shown in Figure 4.18. Click on the text "DC Sweep" to open the DC Sweep window in Figure 4.19, where we specify the nature of the I_{dc} sweep. In the Swept Var. Type field, select Current Source. In the Name field, enter Idc. For Sweep Type, choose Linear. As for the Start Value and End Value, we have chosen –100 mA and +100 mA, respectively. Finally, choosing an Increment of 1 mA will yield 201 data points, which should produce a smooth plot with reasonable resolution. When you have edited your DC Sweep window as shown in Figure 4.19, select OK and Close to return to the *Schematics* page.

At this point, everything is in place. Simulating the circuit (select Simulate from the Analysis menu) should, after a short time, produce the PROBE plot in Figure 4.20. Note that the current I_{dc} is plotted on the *x*-axis and the voltage V_O is on the *y*-axis – just as we requested.

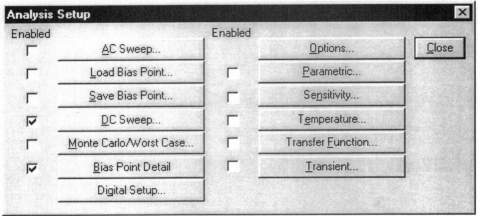

Figure 4.18. A DC Sweep is selected from the Analysis Setup window.

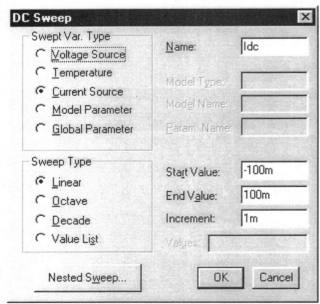

Figure 4.19. The DC Sweep window edited for our particular sweep of I_{dc}.

To extract the required data, activate the cursors by selecting Cursors/Display from the Trace menu. Using the ← → arrows, move the cursor to the point where $V_O = 0$ and read the corresponding I_{dc} value. From the Probe Cursor window in Figure 4.20, the required I_{dc} value is − 80 mA.

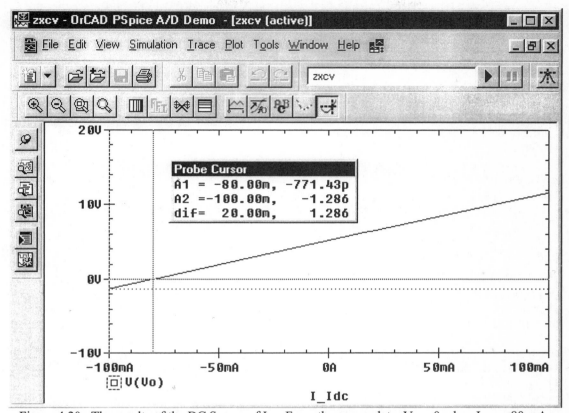

Figure 4.20. The results of the DC Sweep of I_{dc}. From the cursor data, $V_O = 0$ when $I_{dc} = $ - 80 mA.

DC VISUAL TUTORIAL

Throughout this guide we will refer to PSPICE Visual Tutors. These are video files that document the step-by-step processes in circuit creation and simulation. All visual tutors are on the CD-ROM and, since they are executable files, you can run any of them at any time. The first tutor is called DC_TUTOR.EXE. Give it a try.

THE BECA LIBRARY

On the Study Guide CD-ROM you will also find a directory called BECA LIBRARIES which contains four PSPICE library files. These libraries contain custom parts that have been created for your convenience and edification. Text files describing each part can be found in the BECA PART READMES directory on the CD-ROM. These text files can be opened using the PSPICE Text Editor, Notepad or Wordpad. To use the custom parts, you must install the BECA libraries into *Schematics*.

Getting the BECA Library

To get the BECA libraries into PSPICE, copy all four library files (BECA.lib, BECA.slb, BECA.ind and BECA.plb) from the CD-ROM to the subdirectory OrCAD_Demo\PSpice\UserLib.

Including the BECA Library

To use the BECA parts, *Schematics* must be permitted to access them. This process is called "including" the libraries. To include the BECA libraries:

1. Open *Schematics*, go to the Analysis menu and choose Library and Include Files. The dialog box in Figure 4.21 will appear.
2. Using the Browse button, select BECA.lib in the OrCAD_Demo\PSpice\UserLib directory. Click OPEN.
3. Back in the "Library and Include Files" window, click on the Add Library* button. Click OK. Now PSPICE knows where the electrical models for the parts are located.
4. Where are the BECA part symbols? In the Options menu, select Editor Configuration. The window in Figure 4.22 will open.
5. Select Library Settings to open the Library Settings window in Figure 4.23.
6. Using Browse, select the file BECA.slb in the OrCAD_Demo\PSpice\UserLib folder. Click OPEN.
7. Back in "Library Settings", click on Add*, then OK.
8. Back in the "Editor Configuration" window, click OK.

You now have full access to the BECA parts.

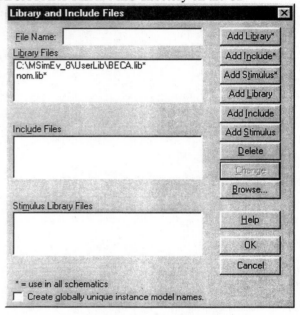

Figure 4.21. Including the BECA.lib library.

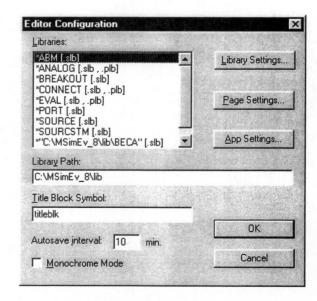

Figure 4.22. The Editor Configuration window.

CUSTOM PARTS

Shown in Figure 4.24 are two parts in the BECA library that may be of some use in dc simulations - the CPL part and the instantaneous wattmeter. The CPL part is a constant power load. As the name implies, it draws constant power regardless of the voltage across it. Details on using the CPL part are given in the file CPL.TXT in the BECA PART READMES directory.

Power can be measured using the instantaneous wattmeter. It has three connections just like its real world counterpart. Look at the READMES file WATTMETER.TXT for instructions on using this part.

In the BECA library, you'll also find a quasi-ideal operational amplifier. Its *Schematics* symbol, equivalent circuit and Attributes box are shown in Figure 4.25. You can specify the op-amp's gain, input and output resistances and supply voltages. VCC is the positive supply and VEE is the negative supply, which have been set to ± 5 V in Figure 4.25c. As in a real op-amp, the output voltage in our quasi-ideal op-amp cannot exceed the supply voltages. Note that the output voltage will be referenced to the ground node in the *Schematics* diagram.

Figure 4.23. The Library Settings window.

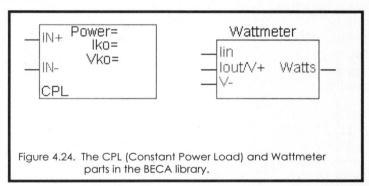

Figure 4.24. The CPL (Constant Power Load) and Wattmeter parts in the BECA library.

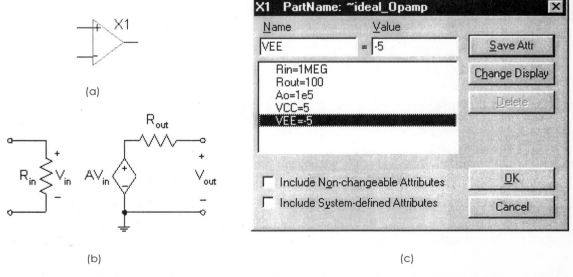

Figure 4.25. The quasi-ideal op-amp in the BECA library: (a) the *Schematics* symbol, (b) the equivalent circuit and (c) the Attributes box.

4.3 Let's use PSPICE to find the voltage, V_O and the power absorbed by the 4-kΩ resistor, in the circuit in Figure 4.26. The corresponding *Schematics* circuit diagram is shown in Figure 4.27. Note that the output node and the wattmeter output have been labeled "Vo" and "Vpower", respectively. To do this, double-click on any wire segment attached to the output node (this opens a simple dialog box) and type the desired node name. Perform the simulation with the SETUP menu set for Bias Point Detail. Selecting the Display Results on Schematic option, the results are shown directly on the schematic. The output voltage is 1.33 V, and the dissipated power is 444.4 μW – exactly V^2/R.

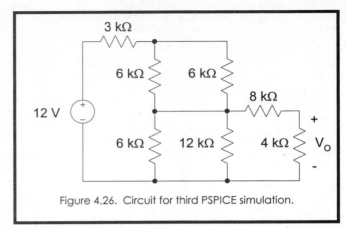

Figure 4.26. Circuit for third PSPICE simulation.

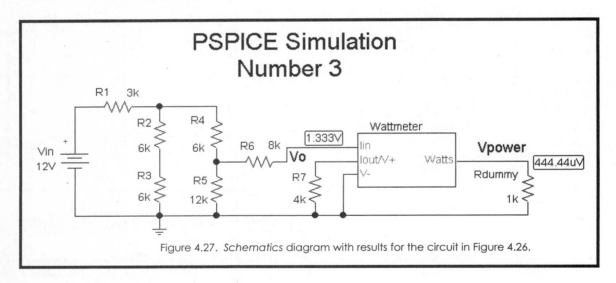

Figure 4.27. *Schematics* diagram with results for the circuit in Figure 4.26.

4.4 Since PSPICE contains parts for all four dependent source types – current-controlled current sources, voltage-controlled current sources, current-controlled voltages and voltage-controlled voltage sources – simulating dependent source circuits is quite easy. All dependent sources are in the ANALOG library. Table 4.1 lists the dependent sources and their PSPICE parts names. Let's simulate the circuit in Figure 4.28a to determine V_O.

TABLE 4.1

PSPICE DEPENDENT SOURCES

Source Type	Part Name
VCVS	E
CCCS	F
VCCS	G
CCVS	H

The *Schematics* diagram for the circuit is shown in Figure 4.28b. To set the gain (multiplicative factor) of a dependent source, double-click on the source to open its Attributes Dialog box, type in the gain and select OK. Simulation results (V_{OUT} = - 24 V) are shown in the schematic using the Display Results on Schematic option.

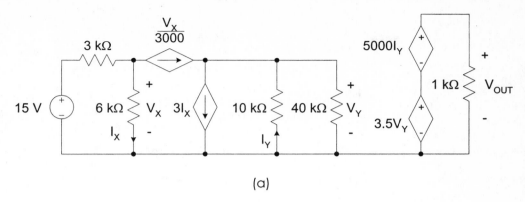

(a)

PSPICE Simulation
Number Four

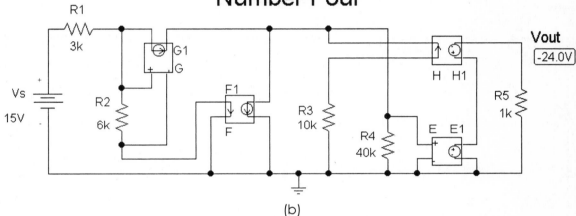

(b)

Figure 4.28. *Schematics* diagrams for, (a) pre-simulation and (b) post-simulation.

4.5 Let's use the quasi-ideal op-amp in the BECA library to simulate a circuit that computes the average value of two voltages.

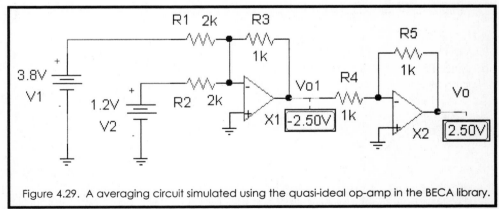

Figure 4.29. A averaging circuit simulated using the quasi-ideal op-amp in the BECA library.

The circuit in Figure 4.29 will perform this task. The output of the first op-amp is

$$V_{O1} = -\frac{R_3}{R_1}V_1 - \frac{R_3}{R_2}V_2$$

For obvious reasons, the first op-amp subcircuit is called a summer. If we make $R_1 = R_2 = 2R_3$, then we have the negative of the average of V_1 and V_2 – namely -2.5 V. The second op-amp subcircuit has a gain of –1, which yields a positive average at V_O.

DOCUMENTATION AND THE README PART

Good engineers always document their work with commentary and informative graphics. *Schematics* provides three methods of documentation. First, terse messages can be added by selecting either Text or Text Box from the Draw menu. Second, graphics can be added by selecting Insert Picture from the Draw menu. And third, extensive commentary can be appended using the README part from the SPECIAL library. Simply add the part to the schematic and double click on it. When its attribute box opens, type in the name of the text file that contains your comments (*filename*.txt). If the file does not exist, you will be prompted to create it. Your text file can be written in any text editor. Most people use Notepad, Wordpad or the PSPICE Text Editor.

Upon finishing your commentary, save it and close the text editor. Now, if at a later date you return to that particular schematic, the README part and the text file reference are included. To open the file, just double-click on the file's name. Note that the *Schematics* file (.sch extension) and the text file must be in the same directories they were in when you specified the README part file name reference. This is easily done if you keep all files relating to a schematic in the same directory.

5. PSPICE simulations of opamp circuits

In the previous chapter, we introduced the quasi-ideal operational amplifier in the BECA library. Let's use this custom part to investigate some simple opamp circuits.

5.1 Let's simulate the circuit in Figure 5.1, which is a simple inverting opamp configuration. The gain of this circuit is

$\dfrac{V0}{Vin} = -\dfrac{R2}{R3} = -\dfrac{50k}{100k} = -0.5$. The attributes for

opamp X1 are given in Figure 4.25(c). Let's vary the value of Vin from -5 V to 5 V and plot the output voltage V0. Activate the Setup window (Setup in the Analysis menu) and select DC Sweep, as shown in Figure 4.18. Click on the text "DC Sweep" to open the DC Sweep window in Figure 5.2, where we specify the nature of the Vin sweep. In the Swept Var. Type field, select Voltage Source. In the Name field, enter Vin. For Sweep Type, choose Linear. As for the Start Value and End Value, we have chosen –5 V and +5 V, respectively. Finally, choosing an Increment of 0.5 V will yield 11 data points. When you have edited your DC Sweep window as shown in Figure 5.2 , select OK and Close to return to the Schematics page.

The results of the DC Sweep of Vin are given in Figure 5.3. Note that this plot verifies the gain expression shown above. For example, when Vin = -2 V, V0 = 1 V, and when Vin = 5 V, V0 = -2.5 V.

Now let's sweep Vin from -15 V to 15 V and plot the results in Figure 5.4. The curve in Figure 5.4 is linear, like the curve in Figure 5.3, over the range -10 V < Vin < 10 V. Outside of this range, the output voltage does not increase with the input voltage. According to the gain

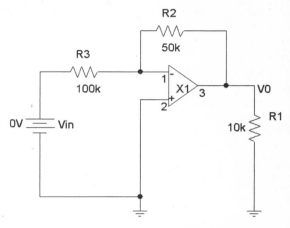

Figure 5.1. Circuit diagram for the first opamp simulation.

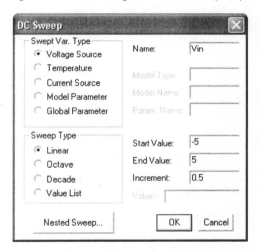

Figure 5.2. DC Sweep window for a sweep of Vin.

calculation above, an input voltage of -12 V should result in an output voltage of 6 V. Note in Figure 5.3 that an input voltage of -12 V results in an output voltage of –5 V. Referring to Figure 4.25(c), the opamp power supplies are set at VCC = 5 V and VEE = -5 V. The output voltage of the opamp cannot exceed the opamp power supply voltage. For $|Vin| > 10$ V, we say that the opamp is in saturation.

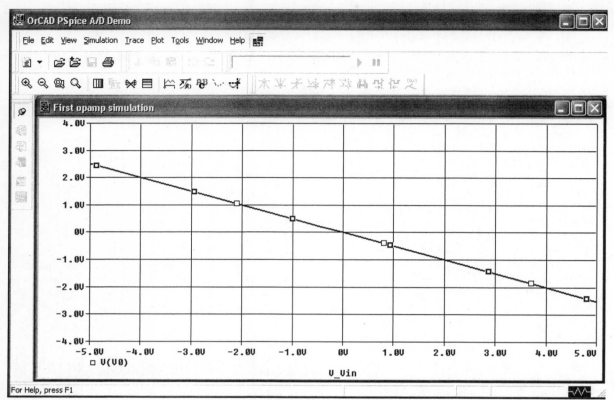

Figure 5.3. The results of the sweep of Vin from -5 V to 5 V.

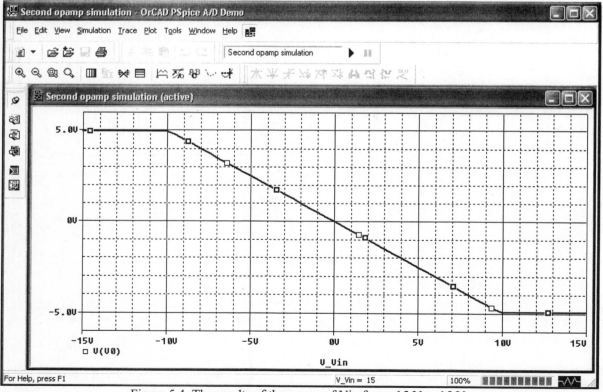

Figure 5.4. The results of the sweep of Vin from -15 V to 15 V.

5.2 Let's now consider the circuit in Figure 5.5. This circuit is very similar to the circuit of Figure 5.1. However, there is one very important difference. In Figure 5.1, the 50 kΩ resistor is connected from the output of the opamp to the negative input terminal, while in Figure 5.5 this resistor is connected to the positive input terminal of the opamp. Let's vary the value of Vin from -15 V to 15 V and plot the output voltage V0. The results of this simulation are given in Figure 5.6. The darker waveform with the diamonds is the output voltage, while the other waveform is the voltage at the input terminals of the opamp. This circuit has limited usefulness as the only two possible output voltages are -5 V or +5 V. The circuit in Figure

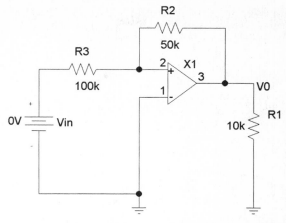

Figure 5.5. Circuit diagram for the third opamp simulation.

5.1 utilizes negative feedback – the output terminal of the opamp is connected through a resistor to the negative input terminal of the opamp. Connecting the output terminal to the positive input terminal results in positive feedback as shown in Figure 5.5. Recall from the model of an ideal opamp that the output voltage is proportional to the voltage difference between the input terminals. Feeding back the output voltage to the negative input terminal maintains this voltage difference near zero to allow linear operation of the opamp. As a result, negative feedback is necessary for the proper operation of nearly all opamp circuits.

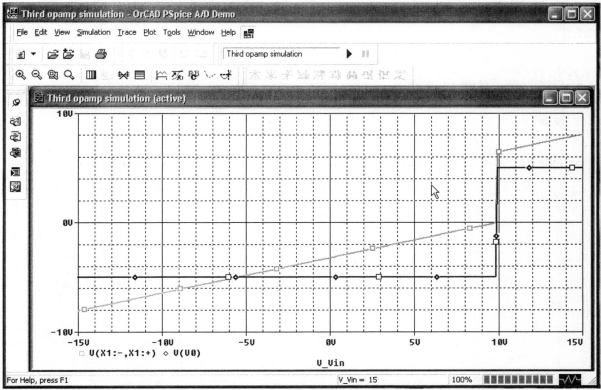

Figure 5.6. The results of the sweep of Vin from -15 V to 15 V for Figure 5.5.

6. Capacitance & Inductance

CAPACITORS

As mentioned in the BECA text and depicted Figure 6.1, capacitors consist of two sheets of metal, called plates, that are separated by a thin sheet of dielectric material. When a voltage is applied between the plates, an electric field must exist inside the dielectric. Since the dielectric is an insulator, no current flows through it and energy is stored in the electric field. Let's compare this situation to that of a resistor. When a voltage is applied to a resistor, an electric field exists in the resistor. Also, in agreement with Ohm's Law, current flows through the resistor. We know the power consumed by the resistor is $P = IV$ and energy usage is the integral of the power. So, an electric field exists in the resistor, the same as in the capacitor, but since current flows, energy is lost as heat rather than stored. The I-V relationship for the capacitor is

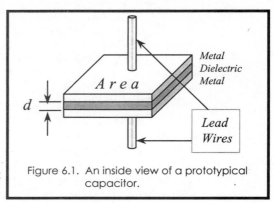

Figure 6.1. An inside view of a prototypical capacitor.

$$i(t) = C\frac{dv(t)}{dt}$$

(6.1)

where $i(t)$ and $v(t)$ obey the passive sign convention and C is the capacitance in Farads (F). Since $i(t)$ is the derivative of charge with respect to time, C can be expressed as,

$$C = \frac{q(t)}{v(t)}$$

(6.2)

where $q(t)$ and $v(t)$ are the charge and voltage on the capacitor respectively. Also, it is shown in the BECA text that the energy stored in the capacitor is

$$w(t) = \frac{C}{2}v^2(t)$$

(6.3)

A simple analogy can be made between electrical energy storage in a capacitor and energy storage in a hydraulic system as seen in Figure 6.2. The switch is analogous to the valve where an open switch (no current) corresponds to a closed valve (no water flow). The charge on the capacitor is analogous to the gallons of water in the vat and the height of the water in the vat is similar to the voltage across the capacitor. The pump speed sets the water flow rate just as the current source sets the charge flow rate. When the valve(switch) is opened(closed) more water(current) flows to the vat(capacitor). This increases the water height(voltage) value. For a given water(current) flow rate, the rate of change in water height(voltage) depends on the area of the vat(capacitor value). So, the capacitance is analogous to

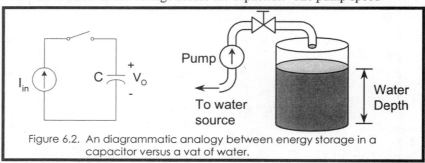

Figure 6.2. An diagrammatic analogy between energy storage in a capacitor versus a vat of water.

the vat's area! Just as a thin vat fills quickly and a wide vat fills slowly, a small capacitor's voltage changes quickly while a larger capacitor's voltage changes slowly.

A capacitor's value is related to its dimensions and materials as defined in Figure 6.1 by the equation,

$$C = \frac{\varepsilon}{d}A$$

(6.4)

where A is the area and ε is the permittivity of the dielectric material. Permittivity, a material constant, quantifies the dielectric's ability to store energy in an electric field. (The permittivity of dielectrics used in capacitor manufacture range from about 0.1 pF/cm to 1000 pF/cm). This can be seen by combining (6.3) and (6.4) as

$$w(t) = \frac{A\varepsilon}{2d} v^2(t) \qquad (6.5)$$

So, for a given voltage and capacitor dimensions, a larger permittivity means more stored energy. Combining (6.2) and (6.4) and solving for ε yields

$$\varepsilon = \frac{d}{A} \frac{q(t)}{v(t)} \qquad (6.6)$$

We see that for a given voltage, more permittivity means more stored charge and thus more stored energy. Since physical circuits operate at specified voltages (i.e. automobile circuits use the 12-V battery) it makes sense to think of higher permittivity yielding more stored charge and more energy per applied volt. From (6.4), capacitors employing dielectrics with large ε values will have large capacitances. For this reason, capacitor manufacturer's are always searching for materials with higher permittivities.

CONSTRUCTION TECHNIQUES

While there are many materials and construction processes for capacitor manufacture, we will focus on the most common: rolled film (paper and plastic), multi-layer (ceramic and mica) and aluminum electrolytic capacitors.

Rolled Construction Capacitors

The capacitor in Figure 6.1 is ill-suited for packaging, especially when C is large. An alternate scheme, called rolled construction, takes narrow strips of dielectric and metal and rolls them up like a jelly roll. This is sketched in Figure 6.3 for a single dielectric layer although many dielectric - metal laminations can be used. To further decrease the size of the capacitor, the metal can be deposited in super-thin layers on either side of the dielectric. This process is growing in popularity. After rolling the strips, lead wires and an external case are added to complete the process.

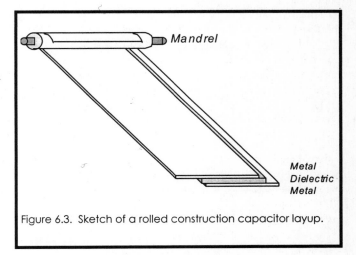

Figure 6.3. Sketch of a rolled construction capacitor layup.

Paper

One of the original dielectrics used in commercial rolled capacitors was paper soaked in oil. While paper has been replaced by plastic dielectrics in most applications, paper caps are still used in high voltage/high power systems.

Plastic

Plastic dielectric capacitors are the most popular capacitors on the market today. They are available in a wide range of capacitance values and working voltages, and they're inexpensive. A partial list of the various plastics used as dielectrics is given in Table 6.1. Specifications for air and paper are included for completeness.

Table 6.1
Partial Listing of Paper and Plastic Dielectrics

Material	Relative Permittivity*	Other Uses
Air	1.0	
Kraft paper	4.4	
Polyester	2.4	wiring insulation
Polyethylene (PET)	3.1	wiring insulation
Polypropolene (PP)	2.2	wiring insulation
Polytetrafluoroethylene (PTFE)	2.1	wiring insulation
*Relative Permittivity is the ratio of ε for material to that of air.		

31

Ceramic Capacitors

A ceramic capacitor looks very much like the prototypical capacitor in Figure 6.1. The dielectric is made of ceramic (aluminum oxide) and the metal layers are deposited. Ceramic capacitors are available in the three configurations shown in Figure 6.4. The lead wires on the button capacitor make it quite versatile. It can be used on printed circuit boards or in a hobbyist's hand-soldered project. The single and multi-layer ceramic capacitors (sometimes called chip capacitors) are packaged for surface mount on PC boards. Having no lead wires make these configurations very small (less than 10 mm on a side).

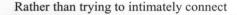

(a)   (b)   (c)

Figure 6.4. Ceramic capacitor configurations. (a) button, (b) chip and (c) multi-level ceramic capacitor.

Aluminum Electrolytic Capacitors

Of the three construction techniques discussed here, electrolytic capacitors have the highest capacitance per cubic centimeter. This is accomplished by drastically reducing the dielectric thickness. Let's begin by explaining the materials inside the electrolytic, which are sketched in Figure 6.5. The metal plates are made of aluminum. One plate is oxidized to form a thin layer (a few 100 nm) of aluminum oxide which serves as the dielectric.

Rather than trying to intimately connect

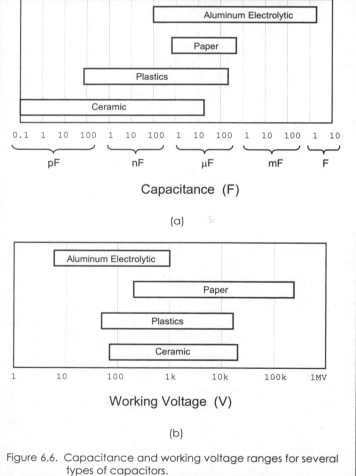

Figure 6.6. Capacitance and working voltage ranges for several types of capacitors.

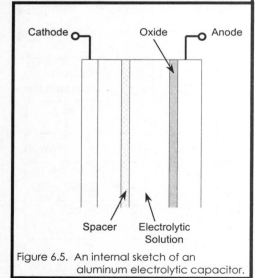

Figure 6.5. An internal sketch of an aluminum electrolytic capacitor.

the second plate to the oxide, an electrolyte solution (usually glycol-based) is added that serves as an intermediate connection. Finally, a porous spacer is added between the plates to keep them apart. To package the capacitor, the plates and spacer are rolled and placed in a cylindrical case.

An unfortunate feature of the electrolytic is that it is sensitive to the polarity of the applied voltage. When the oxidized electrode is positive, the capacitor works great. However, the oxide actually conducts when the voltage polarity reverses! Since the oxidized electrode must be kept positive, it is called the anode and the other plate is the cathode. A mark on the capacitor's case, a positive sign on the anode or a negative sign on the cathode, identifies the terminals.

Now that we've covered the different materials used in capacitor manufacture, we can compare their performance. Capacitor value and working voltage ranges for a variety of dielectrics is given in Figure 6.6.

INDUCTORS

As mentioned in the BECA text, whenever current flows in a wire, a magnetic field is produced. When the wire is arranged in a coil, as shown in Figure 6.7, the magnetic field is focused inside the coil. Thus, energy can be stored in the magnetic field of a coil of wire carrying current. Inductors are little more than coils of wire designed for this purpose.

The *I-V* relationship for the inductor is

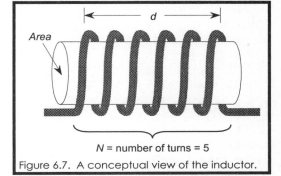

Figure 6.7. A conceptual view of the inductor.

$$v(t) = L \frac{di(t)}{dt} \qquad (6.7)$$

where L is the inductance in Henrys (H). The energy stored in the inductor's magnetic field is

$$w(t) = \frac{L}{2} i^2(t) \qquad (6.8)$$

Inductance is related to the dimensions of the inductor, defined in Figure 6.7, by the equation,

$$L = \frac{AN^2}{d} \mu \qquad (6.9)$$

where N is the number of wraps, or turns, in the coils and μ is the permeability of the material inside the coil. Permeability is a material constant that quantifies how much magnetic field is created per amp of current in the coil. To see how μ affects the inductor performance, combine (6.7) and (6.8) to yield

$$w(t) = \left[\frac{AN^2}{2d} i^2(t) \right] \mu \qquad (6.10)$$

Equation (6.10) indicates that increasing μ will, for a given current, increase the stored energy. For this reason, inductor manufacturers put high permeability materials, called magnetic cores, inside inductor coils.

As seen in Figure 6.8,.an analogy exists between energy storage in the magnetic field of an inductor and energy storage in a flywheel. The energy stored in a spinning flywheel depends on the mass, m, and angular speed, ω (radians/sec.) and is given by

$$w(t) = I\omega^2(t) \qquad (6.11)$$

where I, the moment of inertia, is directly related to mass. Comparing (6.8) and (6.11), it is clear that inductance is analogous to mass. Just as a heavier wheel

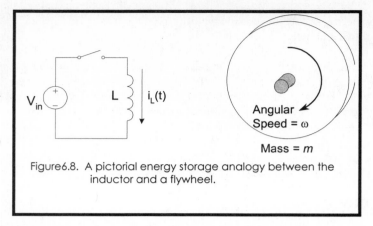

Figure6.8. A pictorial energy storage analogy between the inductor and a flywheel.

stores more energy than a lighter wheel, a larger inductance will store more energy in its magnetic field.

CORES

To increase inductance while maintaining a small component size, manufacturers usually put high permeability materials (called magnetic cores) in their inductors. The most popular cores are iron alloys, powdered iron, molybdenum permalloy powder (MPP) and ferrites. The permeability and useful frequency ranges for each are listed in Table 6.2. Due to its bulk and weight, iron is used mostly in the power industry (60 Hz operation). Powdered iron is exactly that, iron

Table 6.2
Permeability and Useful Frequency Ranges for Several Core Materials

Material	Relative Permeability*	Operating Frequency
Air	1	dc - GHz
Iron Alloys	250 - 2000	dc - 20 kHz
Iron Powder	5 - 80	2 kHz - 100 MHz
MPP	14 - 550	10 kHz - 1 MHz
MnZn Ferrite	750 - 15,000	10 kHz - 2 MHz
NiZn Ferrite	10 - 1500	200 kHz - 100 MHz
* Rel. permeability is normalized μ for air, $4\pi \times 10^{-7}$ H/m		

particles that have been compressed under high pressure. Molybdenum permalloy, an alloy of molybdenum, nickel and steel, is also formed by pressure. Ferrite cores are made of ceramics that contain particles of nickel, zinc, manganese and iron oxide that have been sintered at 2000°C. Due to the way in which they are manufactured, ferrite cores can be made in just about any shape imaginable.

There is a feature of these core materials that is not evident from the above discussion - core saturation. From the BECA text, we know that current and magnetic field are linearly related. This relationship holds for air core inductors. However, for magnetic materials, there is a maximum magnetic field strength that cannot be exceeded regardless of the current. This is shown in Figure 6.9. If the current increases beyond the core saturation limit, the additional energy is dissipated rather than stored in the magnetic field. Since it is important to know the maximum current an inductor can carry before saturation occurs, manufacturers include a maximum current rating specification on magnetic core inductors. Maximum currents range from a few milliamps to tens of amps.

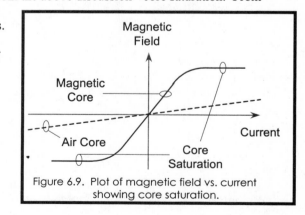

Figure 6.9. Plot of magnetic field vs. current showing core saturation.

7. PSPICE Simulations of Transient Circuits

7.1 Use PSPICE to plot $v_O(t)$ for the circuit in Figure 7.1 over the time interval 0 to 6 µs. Extract the following data.

- The maximum output voltage value.
- The frequency of $v_O(t)$ in Hz.
- Identify the type damping as over-, under-, or critically damped.

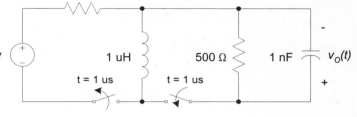

Figure 7.1.

All parts except the inductor, capacitor and the switches have been introduced in Problems 4.1 and 4.2. When getting parts to draw the circuit, use the inductor's part name, L, and the capacitor's part name, C. We will access the switches in a different manner. Select Get New Part from the Draw menu to access the Part Browser in Figure 7.2. Next select Libraries to reveal the Library Browser in Figure 7.3. The switches are in the eval.slb library. Choose it to open the window in Figure 7.4 and scroll down to find parts Sw_tOpen and Sw_tClose. Select and place one of each of the switches.   After wiring, the *Schematics* page should look something like that in Figure 7.5.

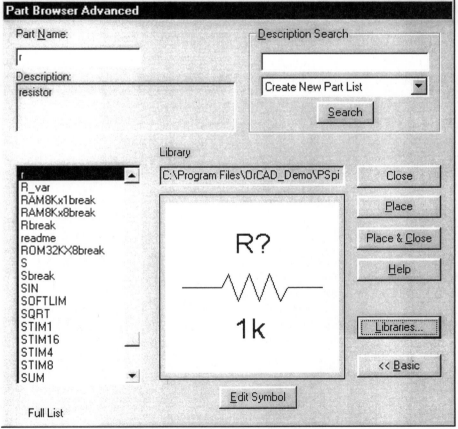

Figure 7.2. The Parts Browser window.

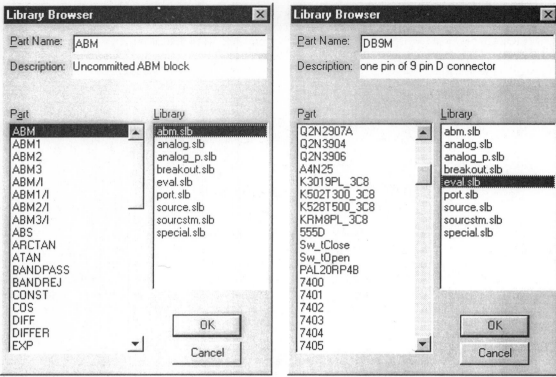

Figure 7.3. The Library Browser window.

Figure 7.4. Locating the switch parts.

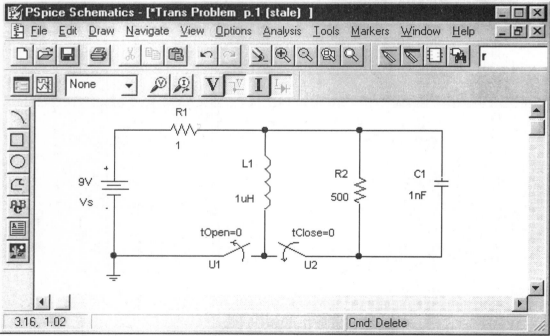

Figure 7.5. The schematic with all parts placed and wired.

Next, we will specify the particulars for the two switches. Double click on the opening switch labeled U1 to view its Attribute window, shown in Figure 7.6.   Fields tOpen and ttran affect the timing of the switch opening. tOpen is the time at which the switch begins to open and ttran is how long it takes the switch to open. Fields Rclosed and Ropen are the effective resistances of the switch as it changes from closed to open. Starting at time tOpen, the switch resistance changes linearly from Rclosed to Ropen over the time interval ttran. It is critical that Rclosed be much less than the other resistances in the circuit and Ropen be much greater than the other resistances. In this way, the affects of the non-ideal switch resistances will be small. For this simulation, the values in Figure 7.6 will do nicely. The other switch, Sw_tClose has the same parameters except tOpen is replaced with parameter tClosed.

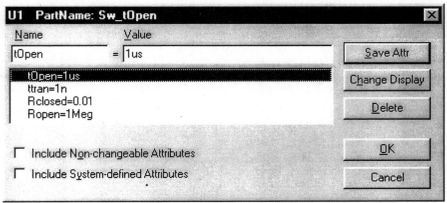

Figure 7.6. The Attribute window for the Sw_tOpen switch.

Adding a voltage marker at $v_O(t)$ would be a convenience. However, the simple voltage marker is referenced to ground whereas $v_O(t)$ is not. Instead, in the Markers menu, select Mark Voltage Differential. Left click at the positive terminal of $v_O(t)$, then left click again at the negative terminal, as seen in Figure 7.7. Now, PROBE will plot the true $v_O(t)$.

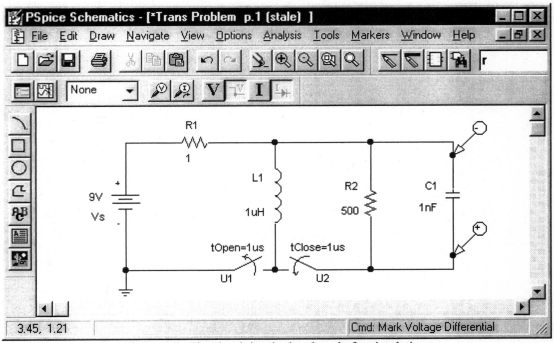

Figure 7.7. The circuit is wired and ready for simulation.

In this simulation, we want to plot $v_O(t)$ versus time. In PSPICE this is called a transient analysis. To specify the transient analysis, select Setup from the Analysis menu. When the Analysis Setup window in Figure 7.8 appears, select Transient. This will open the Transient window in Figure 7.9. The most important field, Final Time, specifies the duration of the simulation. Print Step is useful only when data is being stored in a text file format and is of no interest to us here. No-Print Delay is the amount of time delay before data is printed into a text file or plotted on a PROBE plot. Finally, the Step Ceiling is the maximum interval between data points on the resulting PROBE plot. Of course, a smaller Step Ceiling yields more data points and smoother plots but longer simulation times. Trial and error has shown that selecting Step Ceiling = Final Time/1000 almost always provides a good tradeoff between plot quality and simulation speed. The values in Figure 7.9 are fine for this simulation.

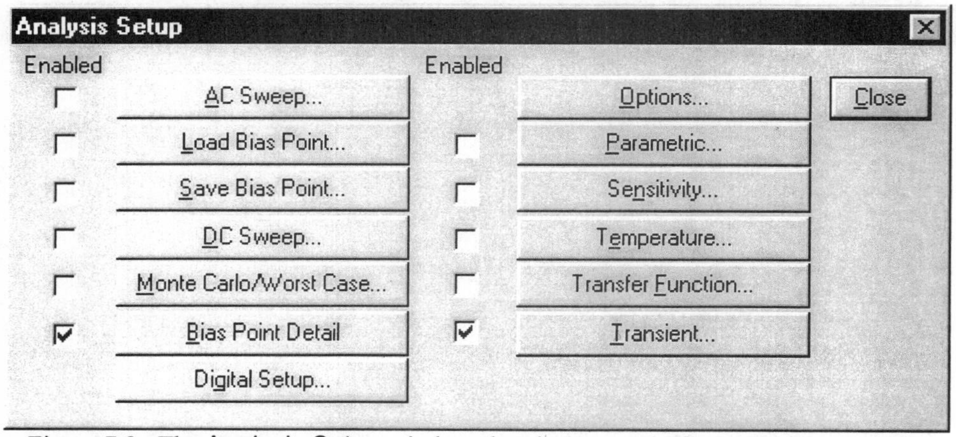

Figure 7.8. The Analysis Setup window showing a request for a transient analysis.

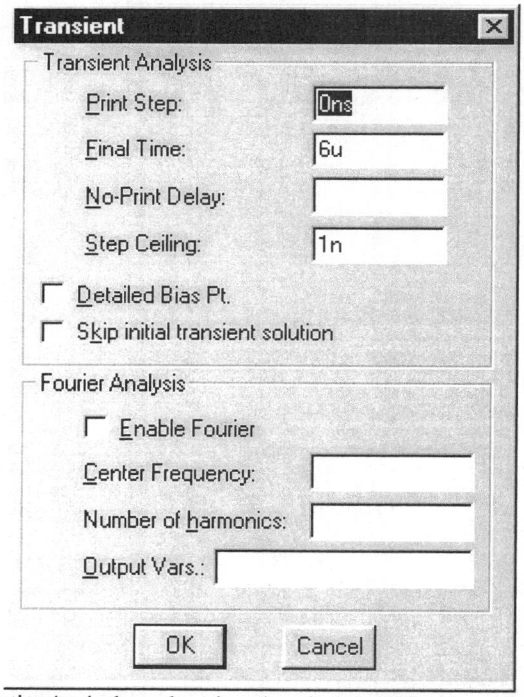

Figure 7.9. The Transient window showing the vital specifications for a transient analysis.

Finally, choose Simulate from the Analysis menu, and PROBE should open, revealing the output voltage in a plot similar to that in Figure 7.10. The maximum voltage value is easily extracted by displaying the cursors (Trace/Cursors/Display) and employing the Max feature (Trace/Cursors/Max). From the resulting Probe Cursor window, shown in Figure 7.11, the maximum output voltage is 266.70 and occurs 49 ns after the switches begin to transition.

In extracting the frequency of the decaying waveform, it would be convenient to zoom in on the plot, isolating just two or three peaks. To zoom in, from the View/Zoom menu, select Area. Back in the PROBE window, click and drag to draw a box around the first three peaks. The result should look something like Figure 7.12. Next, place one cursor at the first peak and the second cursor at the next peak, as shown in Figure 7.13. The time difference in the Probe Cursors window is the period of transient signal. From Figure 7.13, T = 200 ns, and f = 1/T = 5 MHz.

Since the transient waveform is a decaying sinusoid, the circuit is definitely underdamped and $v_O(t)$ is of the form

$$v_o(t) = Ke^{-\sigma t} \cos(\omega_d t + \theta)$$

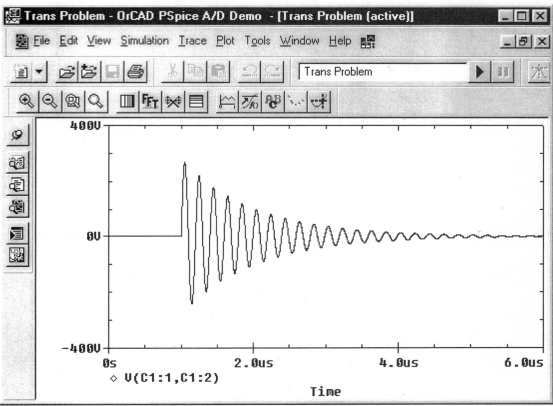

Figure 7.10. The simulation results for $v_O(t)$.

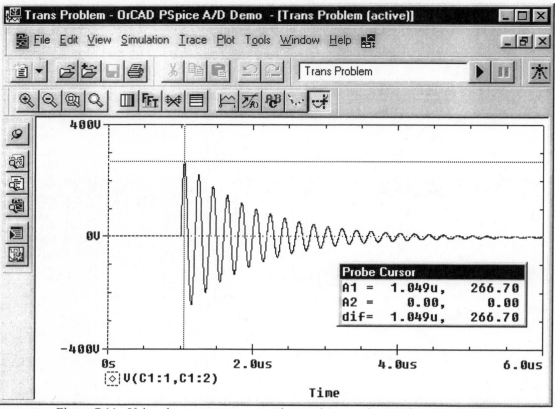

Figure 7.11. Using the cursors to extract the maximum voltage value (266.70 V).

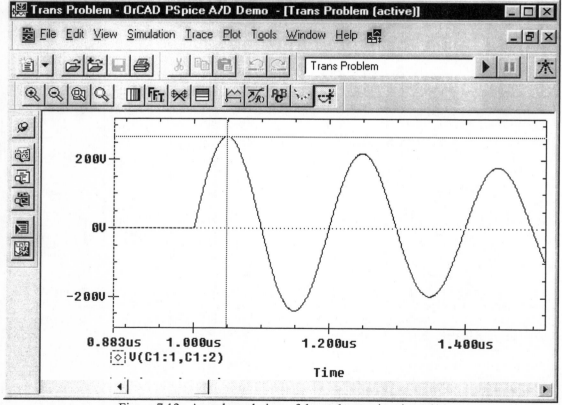

Figure 7.12. An enlarged view of the early transient in $v_o(t)$.

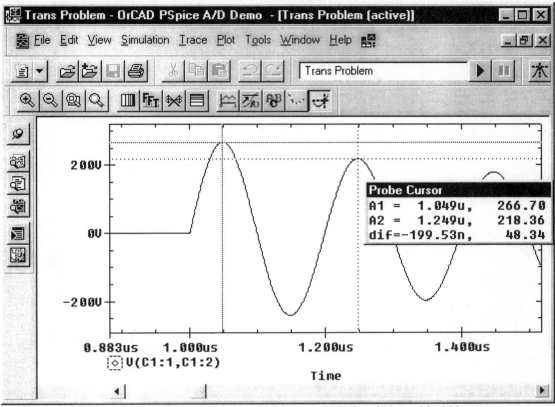

Figure 7.13.   The extracted period of the transient signal is roughly 200 ns.

7.2   Use PSPICE to plot $i_O(t)$ for the circuit in Figure 7.14 over the time interval 0 to 40 μs.   Extract the following data.

- The maximum current value.
- The frequency of $i_O(t)$ in Hz.
- Identify the type of damping as over-, under-, or critically damped.
- The total energy dissipated in the 8-Ω resistor.

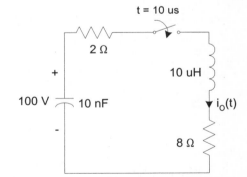

Figure 7.14.   A transient circuit with an initial condition.

The *Schematics* circuit for our basic simulation is shown in Figure 7.15.   To add the initial condition to the capacitor, double-click on the part to open its Attribute window, shown in Figure 7.16 and edit the IC field for the desired value, 100 V.   Also, the switch parameters have been altered from the default values.   The new values are seen in Figure 7.17.

Before we begin the simulation, we must give some thought as to how we will determine the energy dissipated in the 8-Ω resistor.   We know the following

$$p(t) = v(t)i(t) = \frac{v(t)^2}{R} \qquad \text{and} \qquad w(t) = \int p(t)dt$$

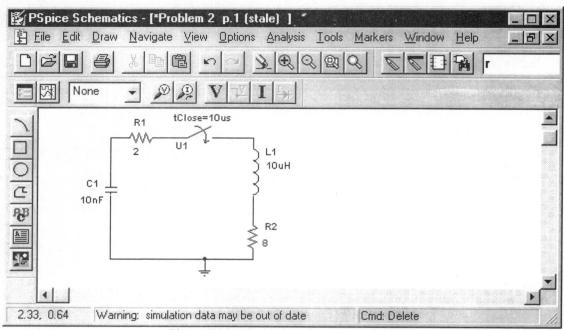

Figure 7.15. The basic *Schematics* circuit.

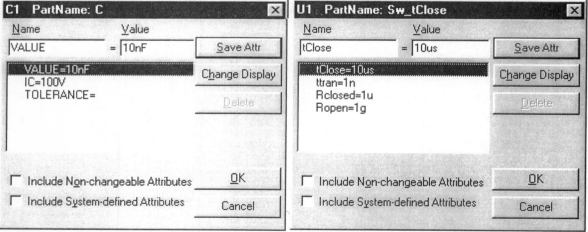

Figure 7.16. An Attribute window for a capacitor reveals the initial condition field – IC.

Figure 7.17. New values for the switch parameters.

Fortunately, *Schematics* has both a multiplier (MULT) and integrator (INTEG) part in the abm.slb library. Adding these parts produces the circuit in Figure 7.18, where the voltage across the 8-Ω resistor is multiplied by itself and then integrated. Double-clicking on the INTEG part reveals its Attribute window, shown in Figure 7.19. GAIN is just a multiplicative prefactor. Since we need to divide $v(t)^2$ by R, the gain has been set to 1/8 = 0.125. The IC option is for adding an initial condition to the integrator output. We want IC set to zero.

In PSPICE, the integrator output cannot be left open circuited. Therefore, a dummy resistor has been added there. Also, the integrator output has been labeled Energy and a Mark Current into Pin marker has be attached to the 8-Ω resistor. The circuit is ready for simulation.

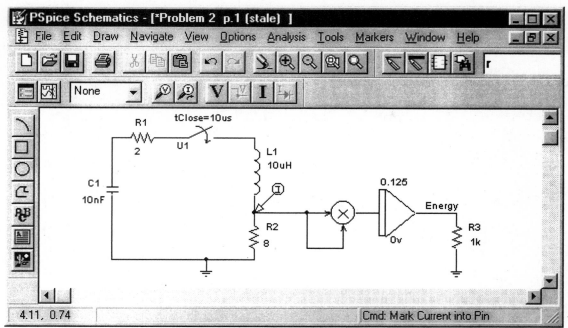

Figure 7.18. The circuit is wired and ready for simulation.

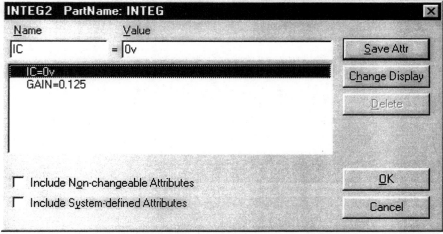

Figure 7.19. The integrator's Attribute window has gain and initial condition fields.

Simulation results are shown in Figure 7.20 where V(Energy) has been added on a second y-axis. To do this, select Add Y Axis from the Plot menu. Then, select Add Trace from the Trace menu. When the Add Trace window appears, just click on V(Energy) and select OK.

To begin extracting data, enable the cursors (Trace/Cursors/Display). We will use cursor 1 to extract the maximum value of $i_O(t)$. First, left click on the $i_O(t)$ trace symbol (the small square). Then, in the Trace/Cursor menu, choose Max. Cursor 1 should jump to the maximum value shown in Figure 7.21, namely, 2.522 A.

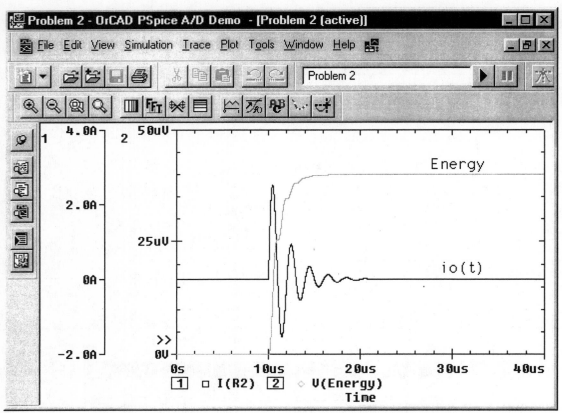

Figure 7.20. Plots of $i_O(t)$ and energy in the 8-Ω resistor.

Probe Cursor		
A1 =	10.45u,	2.522
A2 =	0.00,	100.00n
dif=	10.45u,	2.522

Figure 7.21. The cursor values at the maximum value of $i_O(t)$.

Next, we will extract the frequency of the transient. With cursor 1 still at the maximum value, use the right mouse button to move cursor 2 to the second peak, as shown in Figure 7.22. From the **Probe Cursor** window, the time interval between the cursors is 2.012 µs. This is the period of the decaying sinusoid. Thus, the frequency is

$$f = \frac{1}{T} = \frac{1}{2.012x10^{-6}} = 497 \text{ kHz}$$

As for the type of damping, given the obvious decaying sinusoid waveform, the system is under-damped.

Finally, the energy dissipated in the 8-Ω resistor is calculated by the multiplier and integrator. Simply place either cursor on the V(Energy) trace at any point after transients have subsided. The extracted value is 40 µJ. Let us compare this value to the energy that was stored in the capacitor, which we know to be

$$W_C = \frac{1}{2}CV^2 = \frac{1}{2}(10^{-8})(100)^2 = 50 \text{ µJ}$$

Where has the other 10 µJ gone? It is dissipated in the 2-Ω resistor! Not surprisingly, the ratio of the energy losses equals the ratio of the resistors – 4 to 1. Of course, this is true only because the resistors are in series.

44

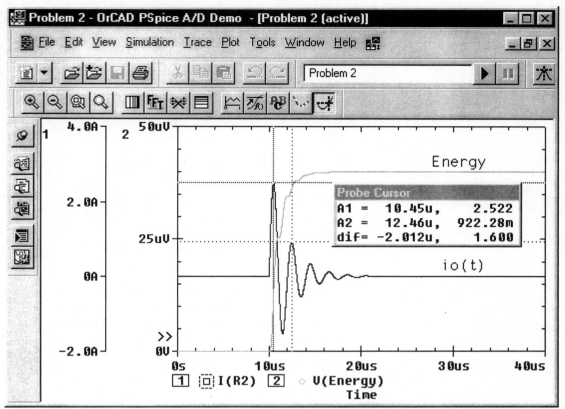

Figure 7.22. Using both cursors, the period of the sinusoidal transient can be extracted.

8. MATLAB Simulations of ac Circuits

8.1 Let us use MATLAB to solve for the equivalent impedance, $\mathbf{Z_S}$, seen by the independent source, $\mathbf{V_S}$, in the circuit in Figure 8.1.

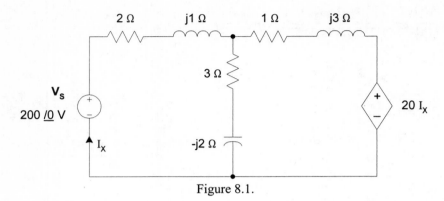

Figure 8.1.

Given the circuit has only two loops, we will use loop analysis, which will yield the loop currents defined in Figure 8.2. Since $\mathbf{I_X} = \mathbf{I_1}$, the impedance in question can be expressed in terms of our loop currents

$$\mathbf{Z_S} = \frac{\mathbf{V_S}}{\mathbf{I_1}}$$

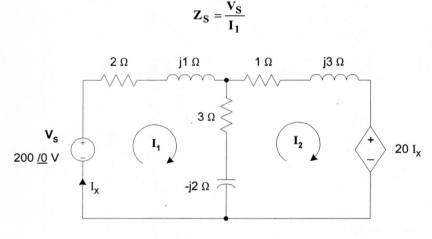

Figure 8.2

Around loop 1, KVL yields

$$200 + j0 = (2 + j1)\mathbf{I_1} + (3 - j2)(\mathbf{I_1} - \mathbf{I_2})$$

or

$$(5 - j1)\mathbf{I_1} - (3 - j2)\mathbf{I_2} = 200 + j0$$

Around loop 2, the KVL equation is

$$20\mathbf{I_X} + (3 - j2)(\mathbf{I_2} - \mathbf{I_1}) + (1 + j3)\mathbf{I_2} = 0$$

or

$$-(3 - j2)\mathbf{I_1} + (4 + j1)\mathbf{I_2} + 20\,\mathbf{I_X} = 0$$

Our final equation is $I_X = I_1$ or, $0 = I_1 - I_X$. In matrix form, we have

$$\begin{bmatrix} 5+j1 & -3+j2 & 0 \\ -3+j2 & 4+j1 & 20 \\ 1 & 0 & -1 \end{bmatrix} \begin{bmatrix} I_1 \\ I_2 \\ I_X \end{bmatrix} = \begin{bmatrix} 200 \\ 0 \\ 0 \end{bmatrix}$$

The loop current matrix can be written as

$$\begin{bmatrix} I_1 \\ I_2 \\ I_X \end{bmatrix} = \begin{bmatrix} 5+j1 & -3+j2 & 0 \\ -3+j2 & 4+j1 & 20 \\ 1 & 0 & -1 \end{bmatrix}^{-1} \begin{bmatrix} 200 \\ 0 \\ 0 \end{bmatrix}$$

This data is entered in MATLAB in the same manner as in the dc cases (see section 3) where Z is the 3x3 impedance matrix and V is the voltage vector.

```
» Z=[5-1j -3+2j 0;-3+2j 4+1j 20;-1 0 1]

    Z =

            5.0000 - 1.0000i  -3.0000 + 2.0000i           0
           -3.0000 + 2.0000i   4.0000 + 1.0000i   20.0000
           -1.0000                            0    1.0000

» V=[200;0;0]

    V =

         200
           0
           0
```

The solution is obtained from the equation

```
» I=inv(Z)*V

    I =

          8.5165 + 5.6572i
        -38.0630 -18.7855i
          8.5165 + 5.6572i
```

where

```
» Zeq=200/I(1)

    Zeq =

       16.2941 -10.8235i
```

Thus, the equivalent impedance seen by V_S is

$$Z_{eq} = 16.2941 - j10.8235 = 19.56 \underline{/-33.59°} \ \Omega$$

8.2 Let us use MATLAB to solve for $\mathbf{V_O}$ and $\mathbf{I_O}$ in the circuit in Figure 8.3.

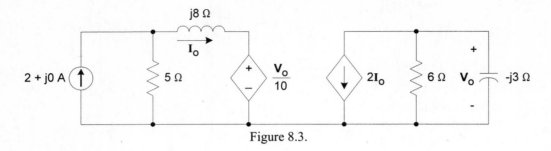

Figure 8.3.

Although the circuit has fewer nodes than loops, we will use loop analysis to solve this circuit. Given the loop current definitions in Figure 8.4, we can write the standard loop equations.

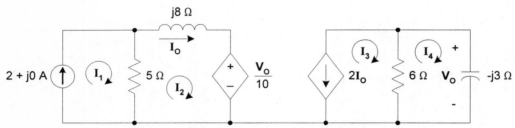

Figure 8.4. Our circuit with loop currents defined.

Loop 1: $\mathbf{I_1} = 2 + j0$ A   or $\mathbf{I_1} = 2$

Loop 2: $5(\mathbf{I_2} - \mathbf{I_1}) + j8\mathbf{I_2} + \mathbf{V_O}/10 = 0$ or $-5\mathbf{I_1} + (5 + j8)\mathbf{I_2} + \mathbf{V_O}/10 = 0$

Loop 3: $\mathbf{I_3} = -2\mathbf{I_O}$ or $\mathbf{I_3} + 2\mathbf{I_O} = 0$

Loop 4: $6(\mathbf{I_4} - \mathbf{I_3}) - j3\mathbf{I_2} = 0$ or $-6\mathbf{I_3} + (6 - j3)\mathbf{I_4} = 0$

$\mathbf{I_O}$: $\mathbf{I_O} = \mathbf{I_2}$ or $\mathbf{I_2} - \mathbf{I_O} = 0$

$\mathbf{V_O}$: $\mathbf{V_O} = -j3\mathbf{I_4}$ or $j3\mathbf{I_4} + \mathbf{V_O} = 0$

These equations can be written in matrix form as

$$
\begin{bmatrix}
1 & 0 & 0 & 0 & 0 & 0 \\
-5 & 5+j8 & 0 & 0 & 0 & 0.1 \\
0 & 0 & 1 & 0 & 2 & 0 \\
0 & 0 & -6 & 6-j3 & 0 & 0 \\
0 & 1 & 0 & 0 & -1 & 0 \\
0 & 0 & 0 & j3 & 0 & 1
\end{bmatrix}
\begin{bmatrix}
\mathbf{I_1} \\
\mathbf{I_2} \\
\mathbf{I_3} \\
\mathbf{I_4} \\
\mathbf{I_O} \\
\mathbf{V_O}
\end{bmatrix}
=
\begin{bmatrix}
2+j0 \\
0 \\
0 \\
0 \\
0 \\
0
\end{bmatrix}
$$

Solving for the unknowns requires solution of the expression

$$\begin{bmatrix} I_1 \\ I_2 \\ I_3 \\ I_4 \\ I_O \\ V_O \end{bmatrix} = \begin{bmatrix} 1 & 0 & 0 & 0 & 0 & 0 \\ -5 & 5+j8 & 0 & 0 & 0 & 0.1 \\ 0 & 0 & 1 & 0 & 2 & 0 \\ 0 & 0 & -6 & 6-j3 & 0 & 0 \\ 0 & 1 & 0 & 0 & -1 & 0 \\ 0 & 0 & 0 & j3 & 0 & 1 \end{bmatrix}^{-1} \begin{bmatrix} 2+j0 \\ 0 \\ 0 \\ 0 \\ 0 \\ 0 \end{bmatrix}$$

In MATLAB, we will let Z be the 6x6 impedance matrix, V will be the 6x1 input vector and I the unknown vector.

```
» Z=[1 0 0 0 0 0;-5 5+8j 0 0 0 0.1;0 0 1 0 2 0;0 0 -6 6-3j 0 0;0 1 0 0 -1 0;0 0 0 3j 0 1]

    Z =
       1          0          0          0          0          0
      -5      5 + 8i         0          0          0         0.1
       0          0          1          0          2          0
       0          0         -6      6 - 3i         0          0
       0          1          0          0         -1          0
       0          0          0      0 + 3i         0          1

» V=[2;0;0;0;0;0]

    V =
         2
         0
         0
         0
         0
         0
```

The unknown currents are then obtained from the equation

```
» I=inv(Z)*V

    I =
        2.0000
        0.5033 - 0.8967i
       -1.0067 + 1.7934i
       -1.5227 + 1.0321i
        0.5033 - 0.8967i
        3.0962 + 4.5681i
```

Thus, I_O and V_O are

$$I_O = 0.5033 - j0.8967 = 1.028 \underline{/-60.70°} \text{ A}$$

$$V_O = 3.0962 + j4.568 = 5.519 \underline{/55.87°} \text{ V}$$

9. PSPICE Simulations of ac Circuits

9.1 Let us simulate the circuit in Figure 9.1 to determine V_O and I_S.

The required *Schematics* diagram including VPRINT1 and IPRINT parts is shown in Figure 9.2. The attributes of the VPRINT1 and IPRINT parts are given in Figure 9.3. From the output file, the requested phasors are given below. The step-by-step procedures used in this simulation are demonstrated in Visual Tutor, ACTUTOR.EXE.

```
 FREQ    Vo-magnitude   Vo-phase
60.0 Hz   0.2308 V       -133.5°


Is-magnitude    Is-phase
 0.3060 A        134.3°
```

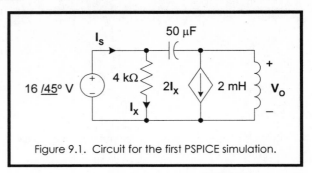

Figure 9.1. Circuit for the first PSPICE simulation.

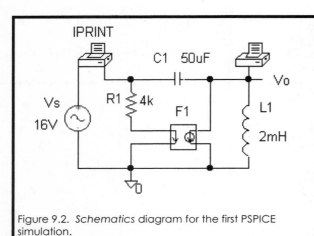

Figure 9.2. *Schematics* diagram for the first PSPICE simulation.

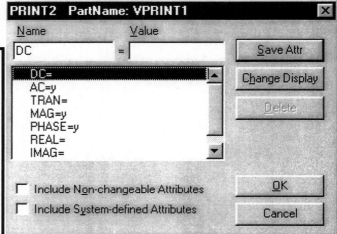

Figure 9.3. The VPRINT1 and IPRINT parts. The VPRINT1 attribute box is set to gather magnitude and phase data from an AC Sweep simulation.

9.2 Let's use PSPICE to find the equivalent impedance of the network in Figure 9.4.

This is a two step process. First, component values cannot be entered as complex numbers. Therefore, we will pick a frequency value for the simulation and calculate the inductor and capacitor values necessary to yield the impedances in the circuit diagram. A convenient simulation frequency is $1/2\pi = 0.159155$ rad./sec. The corresponding component values are given in the *Schematics* diagram in Figure 9.5. The second step is including a current source, I_{TEST}, seen in Figure 9.5, to stimulate the circuit, producing a voltage, V_{TEST}, across I_{TEST}. The equivalent impedance is

$$Z_{eq} = \frac{V_{TEST}}{I_{TEST}}$$

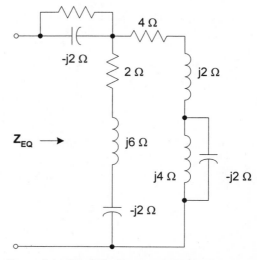

Figure 9.4. Circuit for the second PSPICE simulation.

With foresight, we set $\mathbf{I}_{TEST} = 1\ \underline{/0°}$, then $\mathbf{Z}_{eq} = \mathbf{V}_{TEST}$. The output file results for $\mathbf{V}_{TEST}$ are given below indicating an equivalent impedance of $3.8 + j0.6\ \Omega$.

FREQ (r/s)   0.159

VTEST
    mag.   3.847 V
    phase   8.973 degs
    real   3.800 V
    imag.   0.600 V

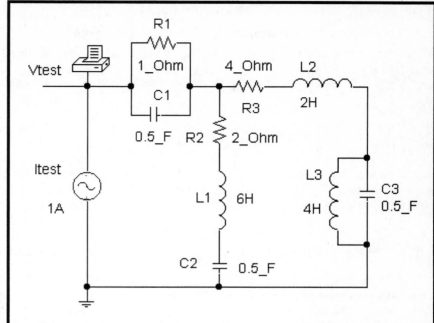

Figure 9.5. *Schematics* diagram for finding the equivalent impedance defined in Figure 9.4.

10. ac Steady-state Power Analysis

The origins of ac steady-state power analysis are found in the power utility industry; right where the BECA text properly focuses our attention. Power levels in kilowatts are quite common. There is however, another application of steady-state power analysis where milliwatts and microvolts are the norm – communication circuitry.

Consider a simple radio set. The radio station broadcasts energy and the set's antenna picks it up. While the station broadcast power might be millions of watts, the voltage on the antenna is typically in the microvolt range. With so little power at the antenna, it is critical that we process the signal efficiently. Thus, engineers design the subcircuits in the radio such that maximum power is transferred between them. Even though the power levels are orders of magnitude lower than those in power utility analysis, the method for optimizing power transfer is the same. Namely, the load impedance must be the complex conjugate of the circuit's output impedance.

11. PSPICE Simulations in ac Steady-state Power Analysis

11.1 Let's simulate the circuit in Figure 11.1 in order to determine the instantaneous and average power absorbed by the resistor, the capacitor and the inductor.

To find instantaneous power, we will use the instantaneous wattmeter in the BECA library. (Proper use and limitations of the wattmeter are explained in the Wattmeter.txt file in the BECA PART READMES folder.) Figure 11.2 shows the *Schematics* circuit where a sinusoidal frequency of 1 Hz has been chosen. Note that the wattmeter output is a voltage which we interpret as power. Simulation results are shown in Figure 11.3 for the power absorbed by each of the passive elements. Note that the power in each element is

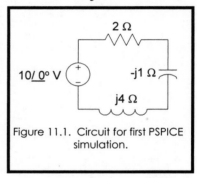

Figure 11.1. Circuit for first PSPICE simulation.

sinusoidal at 2 Hz – twice the excitation frequency. Since the inductor and capacitor power waveforms are centered about zero, they consume zero average power, as expected. However, from the data markers in Figure 11.3, we see that the average power absorbed by the resistor is

$$P_R = \frac{15.37 - 0}{2} = 7.69 \text{ W}$$

Since the *x*-axis variable in Figure 11.3 is time, we know that this is a transient simulation. The voltage source, V1, is a VSIN part from the SOURCE library. Its Attribute box, properly edited for this case is shown in Figure 11.4. The attribute fields bear some explanation. The DC field is the voltage used for dc simulations, the AC field is the phasor magnitude used in AC Sweep simulation and the rest of the fields are used in Transient simulations. Obviously, the VSIN part can be used in any kind of simulation. Next, we perform a single frequency ac simulation at $1/2\pi$ Hz using the circuit in Figure 11.5 where the VPRINT2 and IPRINT parts are set to gather magnitude and phase data. The results are

```
LOOP CURRENT = 2.774A @ -56.31 degrees

COMPONENT     VOLTAGE(V)    PHASE(degrees)
Inductor      11.09           33.69
Capacitor     2.774          -146.3
Resistor      5.547          -56.31
```

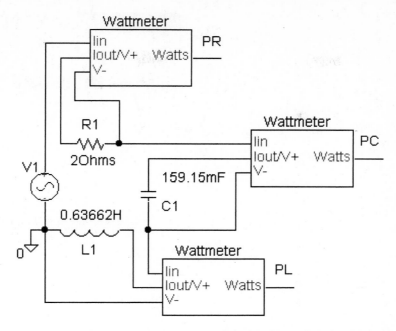

Figure 11.2. The *Schematics* diagram for the instantaneous power simulation.

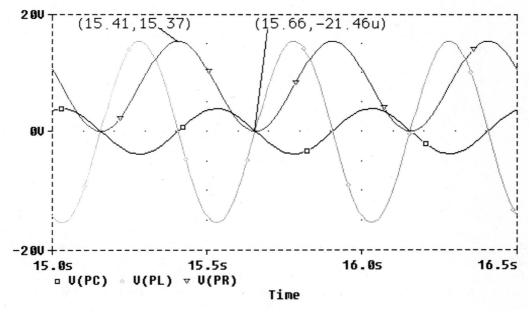

Figure 11.3. The PROBE plot of the absorbed power in the passive elements.

The average power is given by the expression

$$P = \frac{V_M I_M}{2} \cos(\theta_V - \theta_I) \qquad (11.1)$$

Using (11.1) and the simulation results we find that the inductor and capacitor average power is zero while the resistor average power is 7.69 W, exactly the same as in the transient simulation.

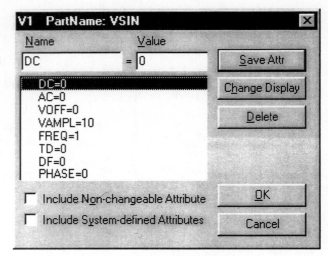

Figure 11.4. The VSIN part Attribute box.

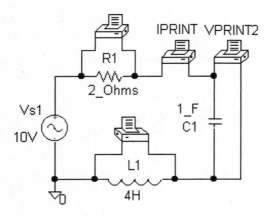

Figure 11.5. *Schematics* diagram for the ac power simulation.

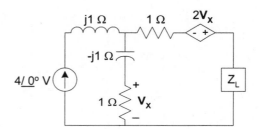

Figure 11.6. Circuit for the maximum power transfer simulation.

11.2. One method for simulating maximum average power transfer is detailed in this example. We will find the load impedance for the circuit in Figure 11.6 that yields maximum power transfer as well as the transferred power.

The simulation procedure is the same as that used in the BECA text. First, find the Thevenin equivalent impedance at the load, Z_{TH}. The *Schematics* circuit in Figure 11.7 should do nicely. We have set the frequency to $1/2\pi$ Hz and the test current source is $1\underline{/0°}$ A. Next, load the circuit with $Z_L = Z_{TH}^*$ as shown in Figure 11.8. Finally, simulate the loaded circuit for the load current and calculate the transferred power. Results for the Z_{TH} simulation are

```
                 TEST RESULTS

         MAG - Vtest     Phase - Vtest

           4.123 V        -14.04 degrees
```

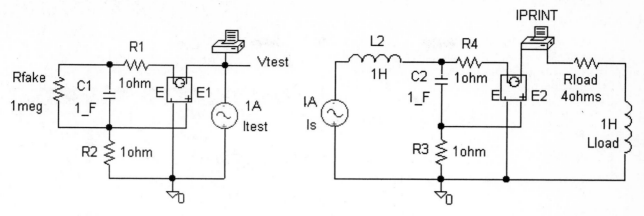

Figure 11.7. *Schematics* circuit for finding Z_TH. Figure 11.8. *Schematics* diagram for finding the transferred power.

Therefore

$$Z_{TH} = \frac{V_{TEST}}{I_{TEST}} = \frac{4.123\underline{/-14.04}}{1\underline{/\ 0}} = 4 - j1 \ \Omega$$

The load impedance for maximum power transfer is

$$Z_{LOAD} = Z_{TH}* = 4 + j1 \ \Omega$$

Simulations for the loaded circuit yield

```
                MAXIMUM POWER RESULTS

        MAG - Iload    Phase - Iload

         1.581 A       -18.44 degrees
```

The maximized load power is therefore,

$$P_{LOAD} = \left(\frac{I_M}{2}\right)^2 R_{LOAD} = \left(\frac{1.581}{2}\right)^2 (4) = 2.5 \ W$$

11.3. Power systems are often specified in wattage, voltage and power factor rather than ohms and henries. To create appropriate *Schematics* diagrams, we must calculate the equivalent load resistances and inductances. For those occasions, we have created a part in the BECA library called LOAD.

Figure 11.9 shows the LOAD part and its Attribute box wherein we can enter the rated voltage, the power factor, the frequency of the simulation and the rated power. PSPICE uses this data to calculate the equivalent resistance and inductance of the load. This means of course that the power factor is always lagging for the LOAD part. Also, errors in the simulated load current will occur if the actual load voltage in the simulation does not equal the rated value entered in the Attribute box. However, since any efficient power delivery system should have minimum transmission losses, these errors should be low.

We will introduce the LOAD part by performing a 60 Hz single-frequency simulation on the circuit in Figure 11.10a to find the line current. The *Schematics* diagram is given in Figure 11.10b where the rated voltage of each LOAD part is set to 220 V rms. From the output file, the simulation results are

```
SIMULATION RESULTS FOR Iline

Magnitude          Phase
318.5 A      -31.09 degrees
```

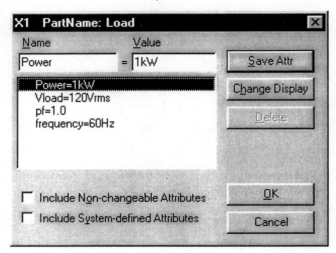

Figure 11.9. The LOAD part and its Attribute box.

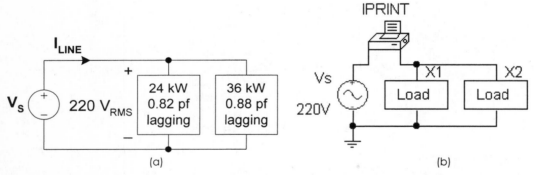

Figure 11.10. Circuits for the third PSPICE simulations, (a) the circuit diagram and (b) the *Schematics* diagram.

12. EXCEL Simulations in ac Steady-state Power Analysis

We wish to use EXCEL to simulate power factor correction for the circuit in Figure 12.1. In particular, the spreadsheet inputs will be the frequency, load voltage in rms, the original power and original power factor. The spreadsheet should then create a plot of the new power factor versus correction capacitor value. Any input can be changed at will, which should be reflected in the plot discussed above. For this example, we will use the following values: $\mathbf{V_S} = 7200\ \underline{/0°}$ V rms, $f = 60$ Hz, $P_{old} = 200$ kW and $pf_{old} = 0.8$ lagging.

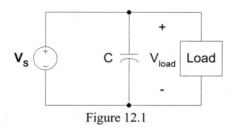

Figure 12.1

First, we determine the old reactive power, Q_{old}, which is given by the expression

$$Q_{old} = P_{old}\left[tan(\theta_{old})\right]$$

where

$$\theta_{old} = arccos(pf_{old})$$

Thus,

$$Q_{old} = P_{old}\left[tan(arccos(pf_{old})\right] \tag{12.1}$$

Next, we must solve for the required correction capacitor value. Adding the capacitor changes only the reactive power to Q_{new}.

$$Q_{new} = Q_{old} + Q_{cap} = Q_{old} - \omega C V_{load}^2$$

Solving for C

$$C = \frac{Q_{old} - Q_{new}}{\omega V_{load}^2} \tag{12.2}$$

From Eq. (12.2), we see that C will be much easier to calculate if we first find Q_{new}, which, from Eq. (12.1), is

$$Q_{new} = P_{old}\left[tan(arccos(pf_{new}))\right] \tag{12.3}$$

This is the approach we will use in EXCEL. First, we will assign values to frequency, load voltage, old power and old power factor. Next, we sweep the new power factor from the original value to unity, determining Q_{new} for each pf_{new} value. Finally, the required capacitor value can be calculated for each Q_{new} value. After the calculations are finished, we can create the required plot.

Upon starting EXCEL, the window in Figure 12.2 should open. We begin with a title for the spreadsheet "Power Factor Correction Spreadsheet" that will appear in the upper left hand corner in 14 point font. To do this, point to cell A1 (first column and first row) and left click once. In the font size drop-down/edit box, choose 14 point font. Next, type the desired text and press ENTER. As shown in Figure 12.3, our title should now be in cell A1.

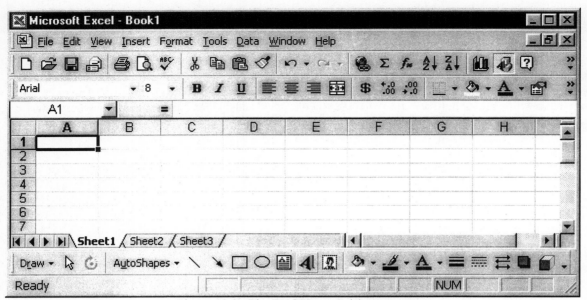

Figure 12.2. The EXCEL spreadsheet program.

Figure 12.3. Adding the main title to our spreadsheet.

Next, we decide which cells will hold the load voltage. Let cell A5 hold the text, "Vload", and cell B5 hold the numerical voltage value and cell C5 holds the units "Vrms". Enter the text "Vload", "7200" and "Vrms" into cells A5, B5 and C5, respectively. Do not include the quotation marks in EXCEL. The results are shown in Figure 12.4. Let's put the frequency right below Vload. In cells A6, B6 and C6, enter the text "Frequency", "60" and "Hz", respectively. Figure 12.5 shows the results thus far.

Let's put the load power in cells F5, G5 and H5 and the old power factor in cells F6 and G6, as seen in Figure 12.6. Now we can do some calculations. Let's input equation (12.1) for Q_{old}. The EXCEL syntax for tangent is TAN and for arccosine, ACOS. Let's put Q_{old} in cells F7, G7 and H7. First put the text "Qold" in cell F7 and "kVAR" in cell H7. To enter Eq. (12.1) into cell G7, first click on cell G7. Then put the cursor inside the text entry field and click once. You should see a blinking cursor in the field. Type the following exactly

=G5*tan(acos(G6))

58

and press ENTER. The equal sign tells EXCEL that you're entering an equation to be evaluated. A value of 150 should appear in cell G7, as seen in Figure 12.7.

We are now ready to create the data for our plot. We need to sweep the new power factor from a value slightly larger than the old power factor to a maximum value of 1. (We will not consider correcting a lagging power factor to a leading power factor since this complicates the math.) Let's put the new power factor values in column A starting at cell A13. In cell A13, enter the text "pf new". In cell A14, enter the following equation

$$=G6+0.005$$

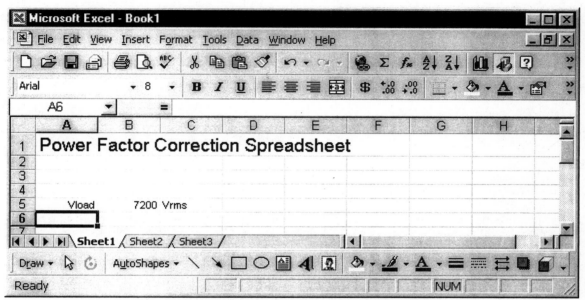

Figure 12.4. Adding the first data with its label and units.

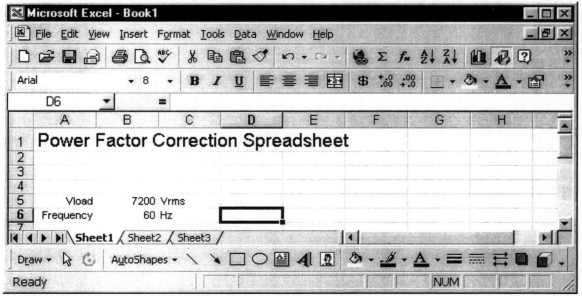

Figure 12.5. Both the load voltage and frequency have been added.

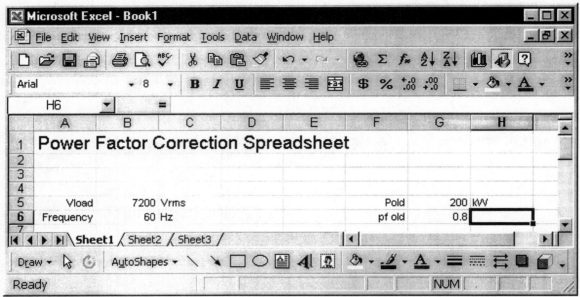

Figure 12.6. All the independent parameters have been added to the spreadsheet.

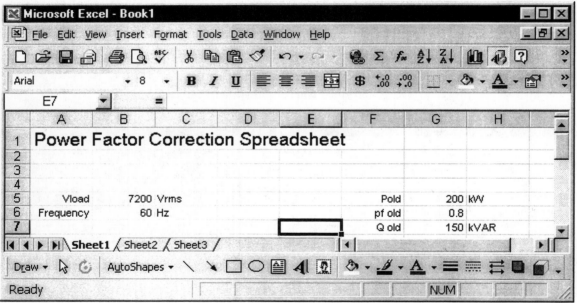

Figure 12.7. All data is now available to create the plot data points.

Our first new power factor value is just 0.005 larger than the old value. Let's fill column A with new power factor values up to 1.00. Select cell A14. In the Edit menu, select **Fill/Series** to open the **Series** window in Figure 12.8 where the fields have been selected to fill column A in 0.005 increments up to 1. Select **OK**, and the results are seen in Figure 12.9, where a long list of new power factor values exist in column A!

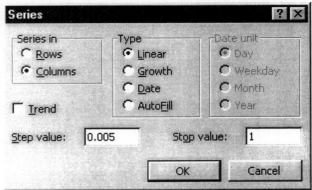

Figure 12.8. The Series window setup for stepping the new power factor up to 1 in 0.005 increments.

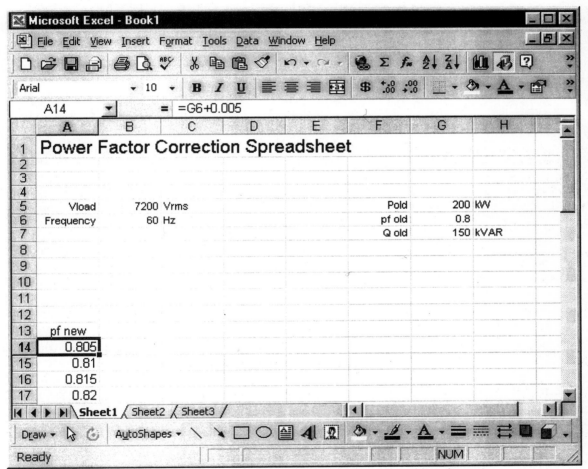

Figure 12.9. The new power factor list has been placed in column A.

Next, we put Q_{new} in column B right next to the new power factor values in column A. Enter the text "Qnew (kVAR)" in cell B13. Then in cell B14, enter the equation

$$=\$G\$5*TAN(ACOS(A14))$$ (12.4)

Now we copy the equation in B14 to the rest of column B. Select B14. A square appears at the lower right corner. Put the cursor on it, left click and hold, then drag the cursor down column B until you reach the last valid row of data. Release the mouse button. You should have a column of Q_{new} values corresponding to the new power factors! The purpose of the dollar signs in Eq. (12.4) should be apparent. The dollar signs force every cell in column B to use P_{old} in cell G5. The work thus far is shown in Figure 12.10.

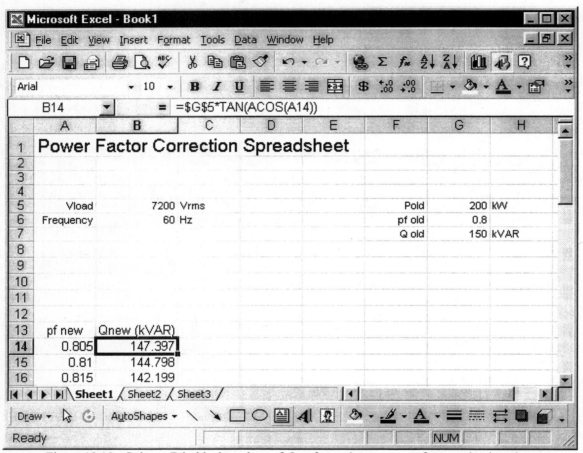

Figure 12.10. Column B holds the values of Q_{new} for each new power factor value in column A.

Now Eq. (12.2) can be handled with ease. In cell C13, enter the text, "C (uF)". In cell C14, enter the equation

$$=1e9*(\$G\$7-B14)/(2*PI()*\$B\$6*\$B\$5^2)$$

where $G7 is Q_{old}, B14 is the first value of Q_{new}, $B6 is the frequency in Hz, $B5 is the load voltage and PI() is EXCEL's quirky way of entering π. A 10^9 prefactor is necessary since Q_{old} and Q_{new} are listed in kVAR and C is listed in μF. As before, copy cell C14 down to the last valid row. Figure 12.11 shows what we've done thus far.

At this point, we could make our plot. There is one simplification which we could use first. In column D, starting at cell D14, enter the equation

$$=A14$$

and copy the cell down to the last valid row. This will simplify the plotting procedure. To plot the data, point the cursor at cell C14, hold, drag down and over until all valid data in columns C and D are selected, then release the mouse button. In the Insert menu, select Chart. The Chart Wizard window in Figure 12.12

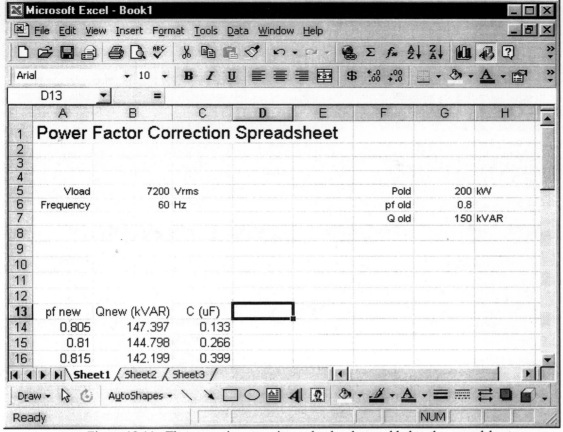

Figure 12.11. The correction capacitor value has been added to the spreadsheet.

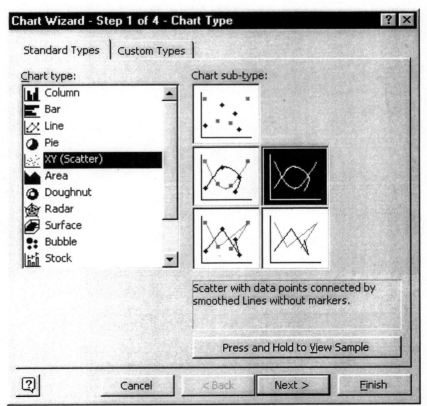

Figure 12.12. Step 1 of 4 in the Chart Wizard allows you to choose the chart type.

will open listing the types of charts you can choose. As shown in the figure, select the smooth XY (scatter) plot without data points and choose Next. This will open Chart Wizard Step 2 of 4. This window should be in good shape. So, choose Next, which opens Step 3 of 4, shown in Figure 12.13. This is where we edit the bulk of the plot cosmetics. Under the Titles tab, you can create titles for the axes and the plot itself. In the Gridlines tab, you choose the gridlines you wish to display. The Legend tab is good for charts that have multiple functions plotted on them. In our case, there is only one function to plot. Thus, we should turn the legend off. The other tabs are not so important. Having edited the fields in the tabs we have mentioned, click Next again to get to Step 4 of 4, which is useful only for very large spreadsheets. Click Next one last time and your chart, shown in Figure 12.14, will appear on the page.

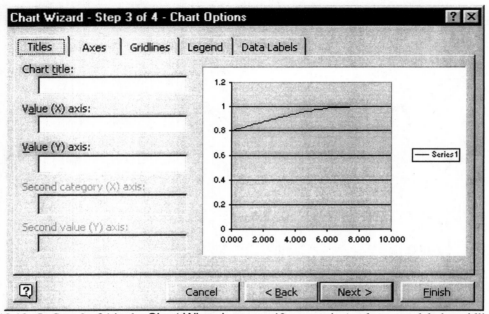

Figure 12.13. In Step 3 of 4 in the Chart Wizard you specify cosmetics such as axes labels, gridlines, etc.

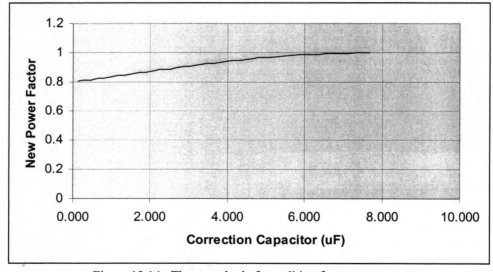

Figure 12.14. The raw plot before editing for appearances.

Usually, you will want to edit the plot for cosmetic reasons. To edit a specific feature such a background color, linewidth, font size, etc., simply double click on the item you wish to edit. For example, Figure 12.15 shows the results of selecting the chart background, where we can change colors and lines. The best advice is to experiment. In particular, look at the x- and y- axes. You can edit font type, font size, scale, tick marks, and so on.

After some editing of our own, the results are shown in Figure 12.16. We see that increasing C improves the power factor, as expected. Furthermore, changing the values for V_{load}, frequency, P_{old} or pf_{old} will be immediately reflected in the spreadsheet data and the plot.

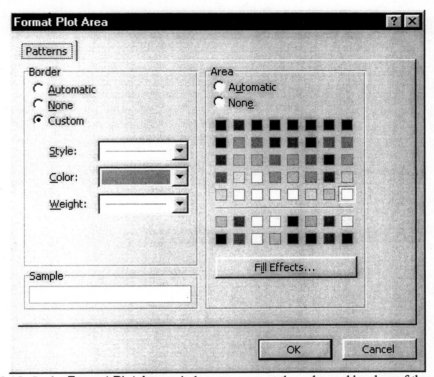

Figure 12.15. In the **Format Plot Area** window you can set the color and borders of the plot area.

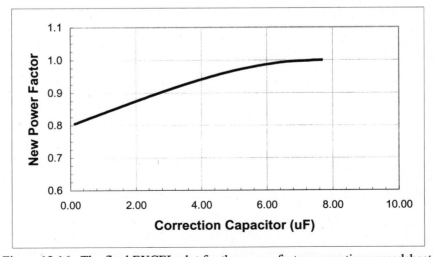

Figure 12.16. The final EXCEL plot for the power factor correction spreadsheet.

13. Magnetically Coupled Networks

A power utility distribution system is an excellent example of a magnetically coupled network. The labyrinth of transformers, substations and cabling that we see every day distributes billions of watts of generated power to millions of customers with high efficiency. We will look at a typical distribution system in detail, discussing the purpose of various substations and pole-mounted equipment.

Another important, yet unseen, example of magnetic coupling between circuits is noise due to magnetic fields. This can be particularly troublesome in instrumentation circuitry. We will consider a prototypical magnetic coupling scenario and simple solutions to the problem.

14. A Typical Power Company Distribution System

Figure 14.1 shows a sketch of the distribution system discussed here. Keep in mind that all voltages are in rms. We start at the generating plant, which can be of the nuclear, hydroelectric or fossil fuel variety. As an example, Figure 14.2 shows an Alabama Power fossil fuel generating plant rated at more than 2 GW. Typical voltages at the generator outputs are around 10 kV. But before leaving the plant, the line voltage is increased to 230 - 500 kV using the transformers in the step-up substation. Of course, the current is decreased by the same amount, which has two benefits. First, lower current means smaller conductors can be used, which lowers costs. Second, lower current leads to lower power loss in the transmission lines. Also, transmission is usually done in a delta connection since three conductors are less expensive than four.

At some point, usually miles from the plant, the first of many transmission substations is encountered. Here, transformers step the line voltage down to 44 kV from 115 kV. Moving down from the transmission sub, we encounter large industries that have their own substations tapping into the power grid at this point. Also downstream from the transmission sub are a number of distribution substations, like the one in Figure 14.3a, where the line-to-neutral voltage is decreased further to 7200 V. The substations you see around your community are distribution subs. Also, most of the power lines are at 7200 V! Businesses such as malls, apartment complexes and restaurants will tie into all three phases, using three transformers as seen in Figure 14.3b to reduce the voltage to usable levels.

For residential customers, there is one final transformer which we will call the step-down center tap transformer that reduces 7200 V to the 120/240 V service at your home. The inner workings of this transformer are depicted in the inset on Figure 14.1. The total secondary voltage is 240 V, but a center tap connection effectively splits the secondary in half. Now we have two sources of 120 V and one source of 240 V. All three secondary conductors are wrapped into one cable, called triplex, shown in Figure 14.3c, that runs to the service entrance at your home. Also, the center tap connection is grounded to the earth both at the power pole and at the service entrance, making it the neutral connection in your home wiring. Once inside your home the conductors are color coded as follows: center tap (neutral) is white, one 120 V line is black and the other is red. A photograph of a 75 kW center-tap transformer is shown in Figure 14.3d.

If you're travelling along a transmission line (44 - 115 kV) you may come across a bank of three regulators, shown in Figure 14.4a. Regulators are variable transformers that automatically adjust their turns ratios to maintain constant line voltage.

As you drive about, you will see plenty of center tap transformers and 7200 V, three-phase wiring lining the streets of your neighborhood. You will also find other pole-mounted equipment such as the 3-ϕ capacitor banks shown in Figure 14.4b. These banks provide power factor correction and a small voltage boost.

Have you ever had the power go out for a few seconds, come back, go out for a few seconds, come back, go out for a minute or so, then come back on and stay on? That's the work of the oil-enclosed, circulating reclosure (OCR) in Figure 14.4c. OCR's are circuit breakers with internal timing circuitry for resetting the breaker. When a short circuit is detected, the OCR opens for a few seconds then recloses. If the short is gone then everything's fine. If not, the OCR opens again. This cycle is usually repeated three times with the last cycle having a longer off time. If the short is still there, the OCR will stay open.

The last piece of pole-mounted equipment we will discuss is the arrestor/switch shown in Figure 14.4d. As the name implies, these serve both as manual switches for safety during repairs and as lightning arrestors. When the switch is open, the bar falls down as seen on the right. In this way, a lineman can spot an open switch from his truck.

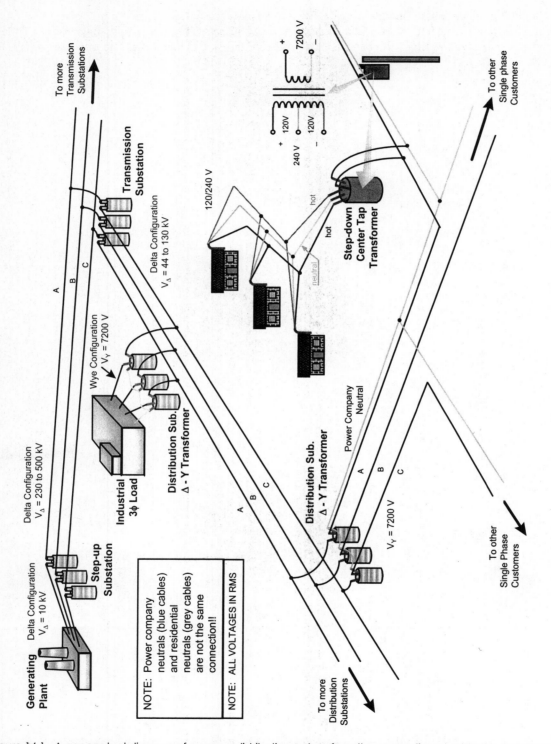

Figure 14.1. A conceptual diagram of a power distribution system from the generating plant to the customer.

Figure 14.2. E.G. Gaston Steam Plant near Wilsonville, Alabama.

Five generating units:

      Units 1 - 4:   265 MW nominal per unit, 280 MW maximum. 100 coal cars per unit per day.
      Unit 5:   860 MW nominal, 960 MW maximum (high pressure unit) 100 coal cars per day.

Total power output: 1.92 GW nominal, 2.08 GW maximum.

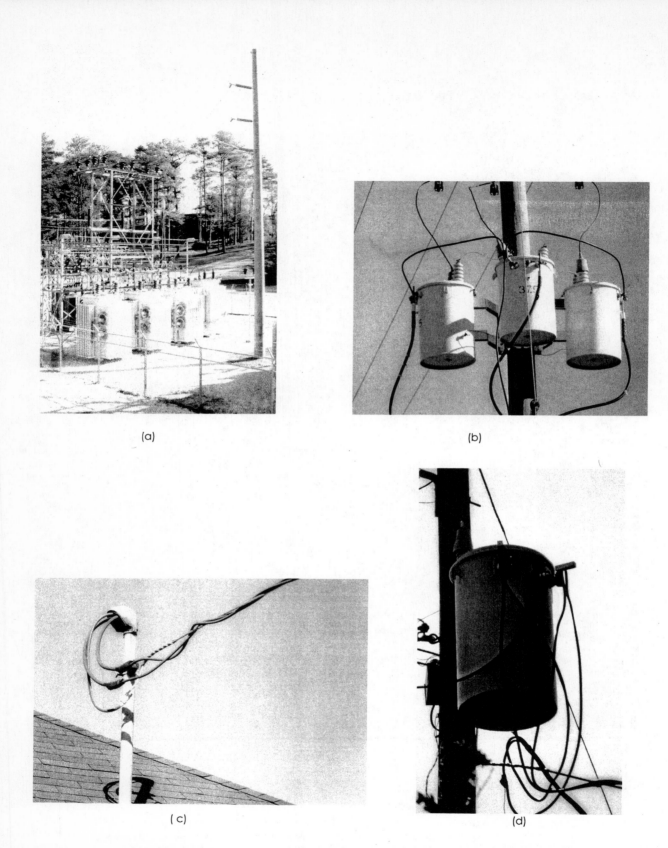

Figure 14.3. A variety of power distribution equipment: (a) Auburn University's substation, (b) three 37.5 kVA center-tap transformers servicing a 16-plex movie theatre, (c) a residential service triplex drop showing triplex wiring, and, (d) a 75 kVA center-tap transformer.

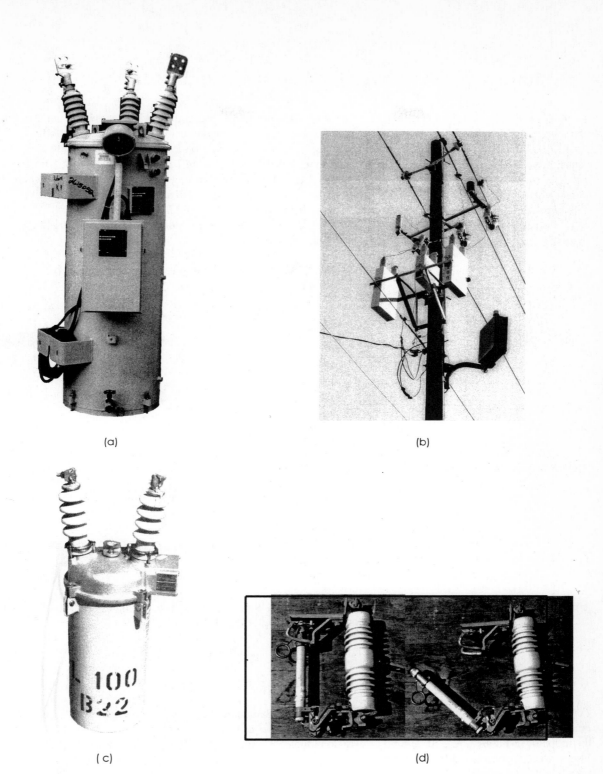

(a)

(b)

(c)

(d)

Figure 14.4. Some additional pole-mounted equipment: (a) automatic regulator (usually found in banks of three), (b) a small bank of power factor capacitors, (c) an OCR (about 2 feet tall), and (d) lightning arrestors/switches in both their closed and opened positions.

15. Simulating Magnetically Coupled Circuits in PSPICE

To model magnetic coupling in PSPICE, we need two things: a way to designate which inductors are coupled and a way to define the dot orientations. Coupled inductors are designated using the K_linear part in the ANALOG library. The K_linear part defines a single coupling coefficient for up to six coupled inductors and is easy to use. Dot conventions are trickier in that the dotted terminal is pin 1 on the inductor. (Pin numbers are discussed in the PSPICE sections in the BECA text.) Here's how you keep it straight: when you get an inductor part, it is placed on *Schematics* horizontally. *For horizontally oriented parts, pin 1 is on the left.* Therefore, the dotted terminal is on the left. If you rotate the inductor, (PSPICE always rotates parts counterclockwise) be cognizant of where pin 1 goes. Should you loose track of pin 1, go to netlist. The inductor's node numbers are always listed pin 1 first, pin 2 second.

15.1 For the coupled inductor circuit in Figure 15.1, use PSPICE to determine the phasor current I_O.

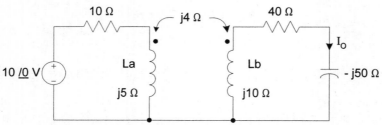

Figure 15.1. A simple coupled inductor circuit.

This simulation has five issues of importance.

1. Since the inductors and capacitors are specified in Ohms, what frequency should we use in our simulation?
2. How do we specify the dot positions on the inductors?
3. How do we set the coupling coefficient between coupled inductors?
4. How do we set up an AC sweep at just one frequency?
5. How do we extract results from our single frequency sweep?

The answer to question 1 is, we can chose any frequency we like! By choosing a frequency, we set all inductor and capacitor values. One of the easiest frequency values to use is $\omega = 1$ r/s. In this way

$$Z_L = \omega L = (1)L = L \quad \text{and} \quad Z_C = 1/(\omega C) = 1/C$$

The resulting inductor and capacitor values are La = 5 H, Lb = 10 H and C = 0.02 F.

To answer question 2, we introduce PSPICE pin numbers. Each part in *Schematics* has pins which are numbered 1, 2, etc. An inductor implicitly has a dot located at pin number 1, which is the pin on the left side of the inductor in Figure 15.2. Should you rotate an inductor, keep track of where pin 1 ends up. For example, in Figure 15.3, both La and Lb were rotated 3 times, putting pin 1 at the top, which, appears to be easier to remember. We will revisit this question after completing the *Schematics* construction.

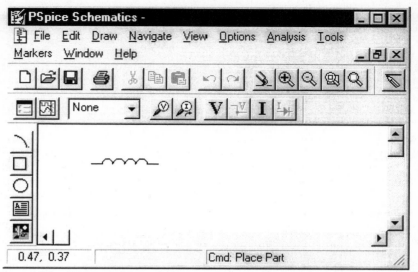

Figure 15.2. Upon getting an inductor, pin 1, the dotted terminal, is on the left.

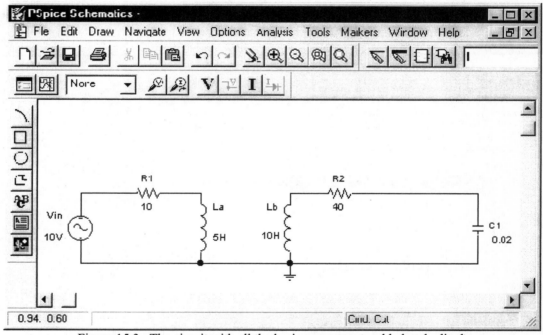

Figure 15.3. The circuit with all the basic components added and edited.

To set the coupling between inductors, we must get a K_linear part from the analog.slb library, which has been added to our *Schematics* page in Figure 15.4. Double click on the K symbol to open its Attribute window shown in Figure 15.5. Up to 6 different inductors can be coupled together using just one K_linear part. We want to couple La and Lb with a mutual inductance of M = 4 H (recall that we chose ω = 1 r/s). Enter La and Lb into the L1 and L2 fields. The coupling coefficient can be determined from the equation

$$k = \frac{M}{\sqrt{L_1 L_2}} = \frac{4}{\sqrt{50}} = 0.565685$$

This value goes into the COUPLING field in Figure 15.5.

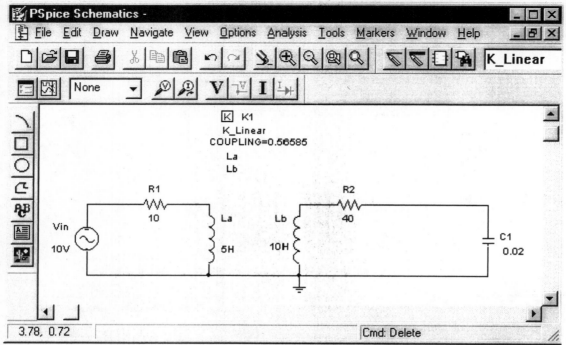

Figure 15.4. The *Schematics* page after adding the K_linear part.

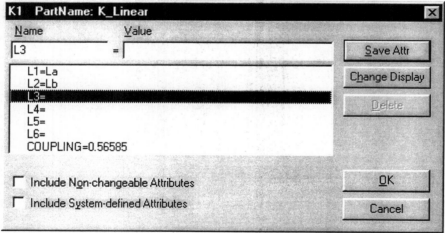

Figure 15.5. The K_linear part's **Attribute** window.

To request a single-frequency AC analysis, select AC Sweep from the Setup window to open the AC Sweep window in Figure 15.6. As shown, simply request a linear sweep of only one point starting and stopping at 0.159165 Hz, which corresponds to 1 r/s.

To answer question 5, we recognize that a plot of I_O versus frequency would be rather silly since there is only one frequency value. It would be much better to print I_O to a file. To this end, we should get a new part from the special.slb library – the IPRINT part. Place the part as shown in Figure 15.7 where I_O enters the unmarked terminal and exits the terminal with the negative mark. Double click on the part to open its Attribute window shown in Figure 15.8, where we have several options. Since we are performing an AC Sweep, we enter YES, or just Y into the AC field. Similarly, entering Y's into the MAG, PHASE, REAL and IMAG fields cause that data to be printed into an output file where it can be easily accessed.

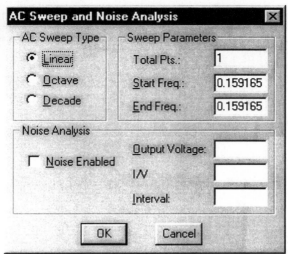

Figure 15.6. Setting up the AC Sweep for a single-frequency analysis at 1 r/s.

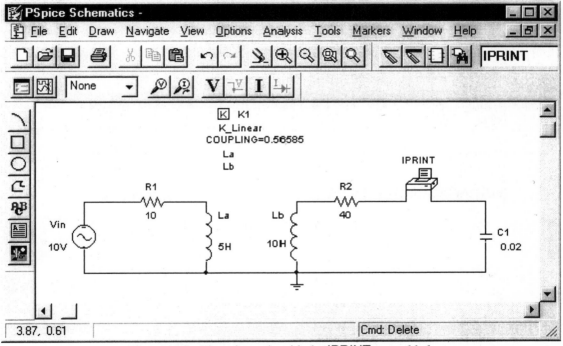

Figure 15.7. The schematic with the IPRINT part added.

The circuit is now ready for simulation. Revisiting question 2, we can check the location of pin 1 on the inductors by accessing the netlist for our circuit. (Choose Examine Netlist from the Analysis menu.) Our netlist is shown in Figure 15.9 where the pin numbers for each part are listed sequentially. So, for La, pin 1 is at node number 1 and pin 2 is at node 0 – the ground node. Similarly, for Lb, pin 1 is at node 5 and pin 2 is at ground. Thus, pin 1 (the dotted terminal) for each inductor is indeed at the top the diagram.

Upon simulating the circuit, (Analysis/Simulate) the PROBE window will open. Let it run its course, and, when it is finished, close it. Back in *Schematics*, select Examine Output from the Analysis menu to open the output file. The data of interest, near the bottom of the file in Figure 15.10, shows the current to be

$$\mathbf{I_O} = -19.07 + j58.77 = 61.78 \underline{/108.0°} \text{ mA}$$

75

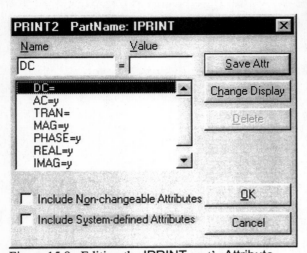

Figure 15.8. Editing the IPRINT part's Attribute window to collect magnitude, phase, real and imaginary components of I_O.

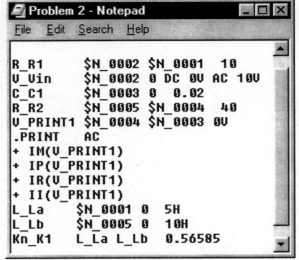

Figure 15.9. The netlist shows the node numbers for each part starting with pin number 1.

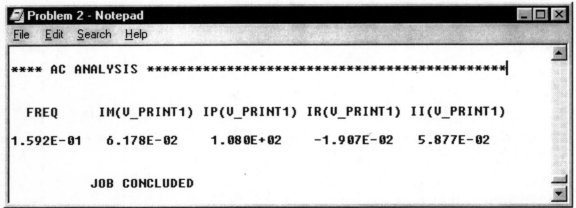

Figure 15.10. The output file results for the IPRINT part yields I_O.

15.2 As an example, let's simulate the magnetically coupled circuit in Figure 15.11 in order to determine the phasor form of the output voltage.

The corresponding *Schematics* diagram in Figure 15.12 includes the K_linear part, which has been edited as shown in Figure 15.13. To set the coupling, we simply enter the names of the coupled inductors and the coupling coefficient into the appropriate fields. The K_linear part can be placed anywhere on the *Schematics* page – it has no connections. When drawing the circuit, we were careful to orient pin 1, and so the dots, at the top of each inductor.

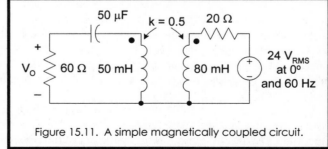

Figure 15.11. A simple magnetically coupled circuit.

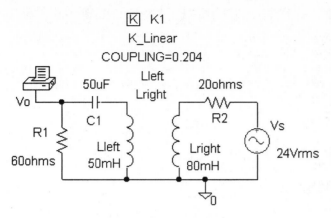

Figure 15.12. Schematics diagram for circuit in Figure 15.11.

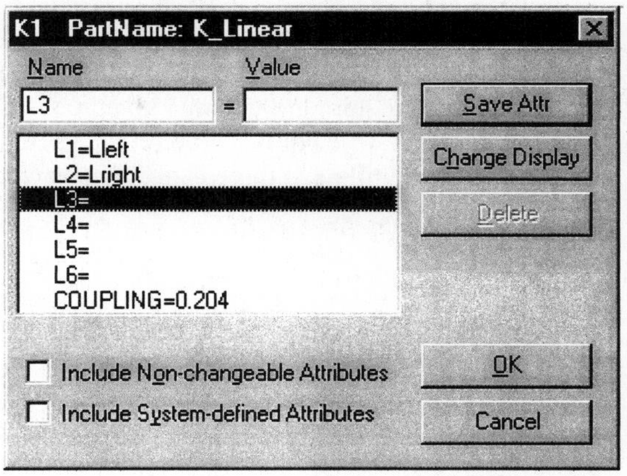

Figure 15.13. The K_Linear Attributes box.

Of course, a single frequency AC Sweep is requested and the phasor data gathered by the VPRINT1 part is in the output file.

```
****     AC ANALYSIS     ****

Vo = 2.779 Volts at 63.48 degrees
```

15.3 In many analyzes, transformers can be treated as ideal without an appreciable loss in accuracy. While there is no ideal transformer in PSPICE, one does exist in the BECA library. Actually, there are two: Ideal_XFMR+ with additive windings and Ideal_XFMR- with subtractive windings. Both transformers, shown in Figure 15.14, have but one editable attribute – the turns ratio (secondary/primary). To demonstrate their use, we will simulate the circuit in Figure 15.15 to determine the impedance at the primary terminals.

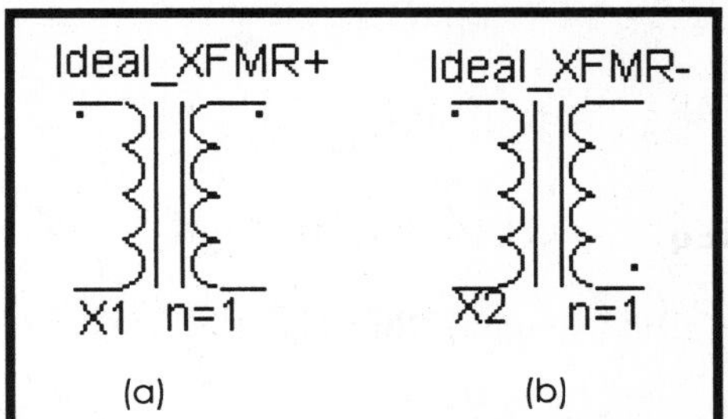

Figure 15.14. The near ideal transformers in the BECA library: (a) additive windings and (b) subtractive windings.

The primary and secondary impedances are related by

$$Z_{PRIM} = \frac{Z_{SEC}}{n^2}$$

Thus, we expect the primary impedance in this case to be 25 Ω. Since the primary current is 1 A, the primary voltage and impedance have the same numeric value. From the output file, the primary impedance is dead on.

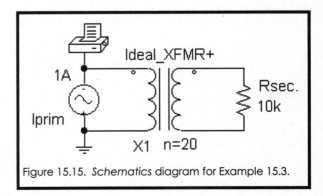

Figure 15.15. *Schematics* diagram for Example 15.3.

```
****     AC ANALYSIS     ****

Zprimary = 25.00 Ohms at 0.0
```

degrees

16. EXCEL Demonstration for ac Steady-state Power Analysis

On the CD-ROM, you will find the EXCEL file Power Factor Correction.XLS. This spreadsheet can calculate the capacitance required to obtain a given power factor. The file was written to accept and output data for a single-phase wye-wye equivalent circuit. So, to use the file for three-phase calculations, divide the total power by 3. Also, you will need three capacitors, each equal to the calculated value. An example of the file is shown in Figure 16.1.

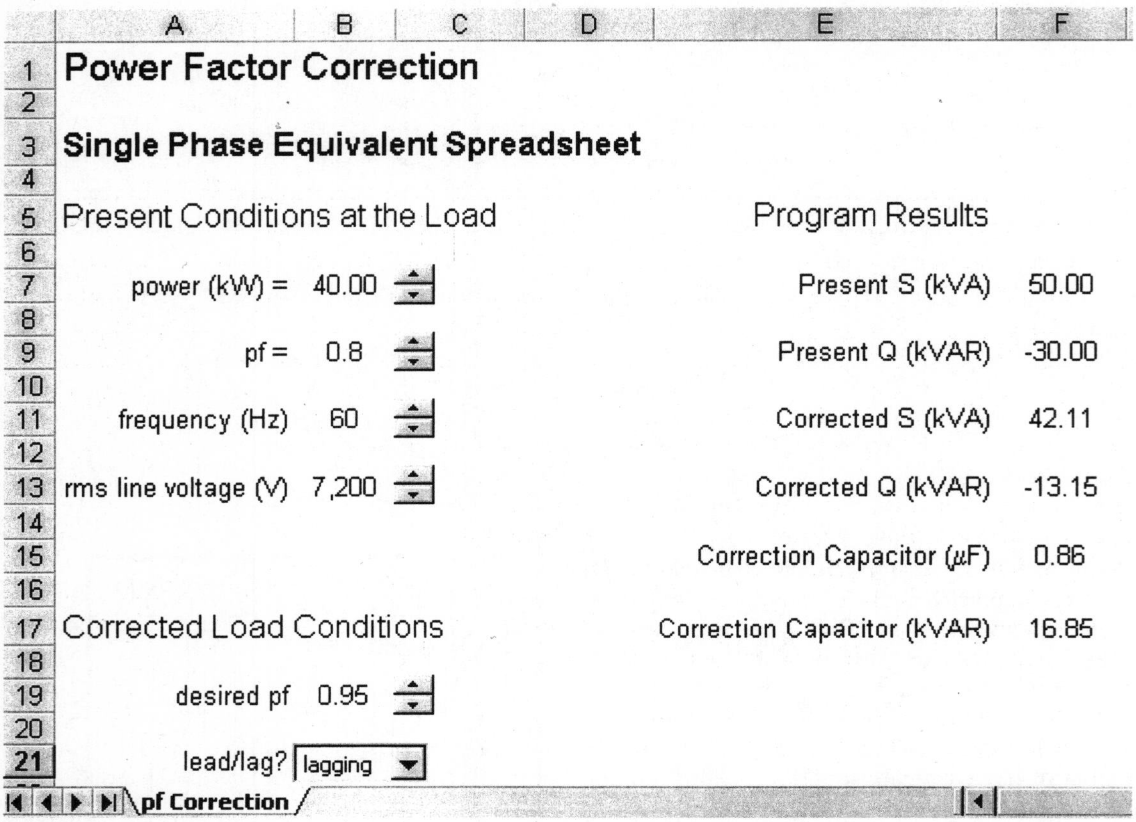

	A	B	C	D	E	F
1	**Power Factor Correction**					
2						
3	**Single Phase Equivalent Spreadsheet**					
4						
5	Present Conditions at the Load				Program Results	
6						
7	power (kW) =	40.00			Present S (kVA)	50.00
8						
9	pf =	0.8			Present Q (kVAR)	-30.00
10						
11	frequency (Hz)	60			Corrected S (kVA)	42.11
12						
13	rms line voltage (V)	7,200			Corrected Q (kVAR)	-13.15
14						
15					Correction Capacitor (μF)	0.86
16						
17	Corrected Load Conditions				Correction Capacitor (kVAR)	16.85
18						
19	desired pf	0.95				
20						
21	lead/lag?	lagging				

pf Correction

Figure 16.1. An example of the file Power Factor Correction.XLS.

17. Polyphase Circuits

For some reason, three-phase circuit analysis gives some students a lot of trouble. The jargon may be largely to blame. To avoid confusion, it is good to have a few simple definitions in hand. It is even better if those definitions do not depend on the topology – delta vs. wye. The power company has this perspective. They are concerned with power delivery to customers. Whether a motor is wye or delta connected is of little interest. It is not surprising then that the terms used to define power delivery, listed in Table 17.1 are topology independent. Notice that these terms actually define the power line – its voltage, current and impedance. The terms in Table 17.1 are well worth memorizing as they serve as a point of reference for other three-phase terminology.

Table 17.1

Topology-independent Terms in 3-ϕ Jargon

Term	Description	BECA Equivalent
Line impedance	Resistance and inductance of the actual power lines	Z_{line}
Line current	Current in the actual power lines	$\mathbf{IaA}$, etc.
Line voltage	The line-to-line voltage	$\mathbf{V_{AB}}$, etc.
Line-to-Neutral Voltage	The voltage from a power line to the neutral	$\mathbf{V_{AN}}$, etc.

When discussing the voltages and currents in a particular load or source, the topology affects the meaning of the terms we use. In particular, the terms *phase current* and *phase voltage* are completely dependent on the connection scheme. Do not confuse them with the terms in Table 17.1 – they are not synonymous. However, if we look at these two terms from the perspective of the load rather than the line, we find they are similar. The phase voltage is the voltage across any one of the three load components, while the phase current is the current through a load component. This is shown in Figure 17.1 for both delta and wye connections.

Notice that the line current (the actual transmission line current), is the same as the wye-connected phase current. As a result, the line impedances are in series with the load components. This is the motivation for converting balanced loads and sources to a wye-wye format. It allows us to combine source, line and load particulars into a single-phase analysis whose solution yields the currents in the physical transmission lines. So, think wye-wye.

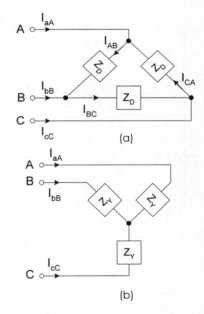

Figure 17.1. Generic (a) delta and (b) wye load connections.

18. Simulating Three-Phase Circuits in PSPICE

Many power system loads are specified in rated power, voltage and power factor rather than resistance and inductance. This makes circuit simulators like PSPICE and EWB less than optimum for power system simulations. In fact, we will not perform any EWB simulations for 3ϕ circuits. Another drawback is that drawing balanced three-phase loads in *Schematics* is a bit redundant since the only difference between phases is a 120° phase shift. To speed things along, balanced loads and sources for both wye and delta configurations have been created and included in the BECA library. The names and attributes of the parts are listed as follows.

Six balanced 3-φ parts have been created for this study manual. They are shown in Table 18.1 and included in the BECA library for your use. While the editable attributes of each part are listed in Table 18.1, detailed descriptions of each part are in the READMES folder on the CD-ROM. One point of interest regarding the loads: the part attributes are used to determine effective resistance and inductance/capacitance of the load. If the actual load voltage and the rated voltage attribute are not equal, errors will result. This will be discussed in simulation 19.3 in Chapter 19.

Table 18.1

Parts List And Attributes For The BECA Library 3-φ Parts

Part Name	Function	Attributes
V_delta	Delta connect. Source	Vab = Line-to-line voltage
		Phaseab = phase of V_{ab} in degrees
		Sequence = +1 for abc; -1 for cba
		Freq = Source frequency in Hz
V_wye	Wye connect. Source	Van = Line-to-neutral voltage
		Phasean = phase of V_{an} in degrees
		Sequence = +1 for abc; -1 for cba
		Freq = Source frequency in Hz
Delta_Load _Inductive	Lagging Δ Load	Vab = Line to line voltage
		VA = Total load volt-ampere rating
		pf = Power factor
		frequency = Source frequency
Delta_Load _Capacitive	Leading Δ Load	Vab = Line to line voltage
		VA = Total load volt-ampere rating
		pf = Power factor
		frequency = Source frequency
Wye_Load _Inductive	Lagging Y Load	Van = Line to neutral voltage
		VA = Total load volt-ampere rating
		pf = Power factor
		frequency = Source frequency
Wye_Load _Capacitive	Leading Y Load	Van = Line to neutral voltage
		VA = Total load volt-ampere rating
		pf = Power factor
		frequency = Source frequency

19. PSPICE Simulations of Three-phase Circuits

In this chapter, we will perform four simulations. In simulation 19.1, the loads are given in resistance and inductance. In simulation 19.2, we will make use of the custom parts mentioned above. In simulation 19.3, we will investigate shortcomings of our custom loads. Finally, in simulation 19.4, we will demonstrate how to find the required capacitance values for power factor correction applications.

19.1 A particular 60 Hz balanced delta-wye system has a 220-V rms, abc sequence source with $\underline{/\,V_{ab}} = 20°$. The load impedance per phase is $8 + j6\ \Omega$ and the line impedance is $0.4 + j0.8\ \Omega$. Let's use PSPICE to find the line currents and phase voltages at the load in phasor form.

The circuit in Figure19.1 will do nicely. We have used the V_delta part in the BECA library for the balanced source, and, while the plethora of VPRINT1 and IPRINT parts is visually stunning, there are only three of each – one for each phase. Also, each VPRINT1 and IPRINT part has been set to acquire magnitude and phase data.

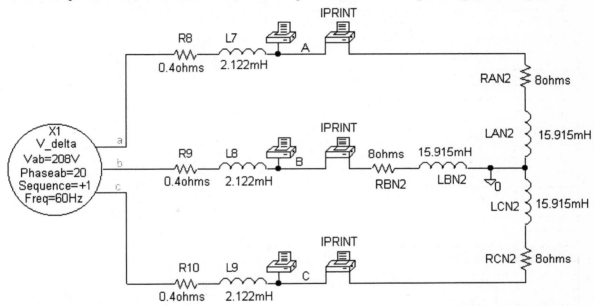

Figure 19.1. The *Schematics* diagram for simulation 19.1 employing the V_delta source and VPRINT1 / IPRINT data gathering parts.

The simulation results are listed below.

**** AC ANALYSIS ****

PHASE VOLTAGES	MAGNITUDE(RMS)	PHASE	LINE CURRENTS	MAGNITUDE(RMS)	PHASE
Van	1.111E+02	-1.212E+01	IaA	1.111E+01	-4.899E+01
Vbn	1.111E+02	-1.321E+02	IbB	1.111E+01	-1.690E+02
Vcn	1.111E+02	1.079E+02	IcC	1.111E+01	7.101E+01

Note that the phase voltages (line-to-neutral for the wye-load) are equal in magnitude and 120° out of phase. The same can be said of the line currents.

19.2 Two industrial plants receive their power from the same substation where the line voltage is 4.6 kV rms and the phase sequence is *abc*. Plant 1 is rated at 300 kVA and pf = 0.8 lagging, while plant 2 is rated at 350 kVA at 0.64 lagging. The source and loads are balanced. Let's find the line currents at the source and the line currents into each plant.

Since no information is given regarding delta or wye connections, we will use a delta-delta connection. In this way, we do not have to convert line voltage to line-to-neutral voltage. Furthermore, as seen in Figure 19.2, we will use the V_delta and Delta_Load_Inductive parts to model the system. Even though the delta parts in the diagram have no neutral connection, PSPICE still requires a grounded node somewhere. We arbitrarily chose to put it at phase C. Also, since we know the phase angle relationships between the phases, we will use IPRINT parts to extract the line currents in the A phase only.

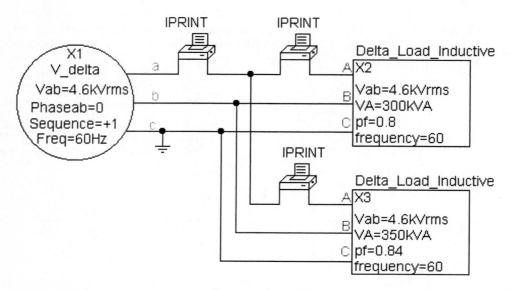

Figure 19.2. *Schematics* circuit for Simulation 19.2 using the delta load and source parts in the BECA library.

The simulation results are listed below. We can check the simulation results by adding the load currents. They should equal the source current.

$$I_{LOAD1} + I_{LOAD2} = 37.65 \underline{/-66.87°} + 43.93 \underline{/-62.83°} = 14.79 + 20.03 - j(34.62 + 39.09) = 81.53 \underline{/-64.71°} \quad Q.E.D.$$

**** AC ANALYSIS ****

PHASE A LINE CURRENTS	MAGNITUDE (A RMS)	PHASE (DEGREES)
Source	81.53	-64.71
Load 1	37.65	-66.87
Load 2	43.93	-62.86

19.3 As mentioned earlier, the attributes of a BECA 3-ϕ load are used to calculate equivalent load resistance and inductance/capacitance values which are sent to PSPICE as circuit elements. Thus, the circuit PSPICE eventually solves looks very similar to Figure 19.1. There is a problem however when the actual load voltage is not the same as the attribute value.

As an example, consider a 60-Hz, balanced wye-wye connected system where the source voltage is known to be 208 V rms and the line resistance and inductance are 0.1 Ω and 1 mH, respectively. Also, the load is *supposed* to operate at 208 V and 18 kVA at 0.75 power factor lagging. We know that with line impedance included in the simulation, it is impossible for both the source and load voltages to be 208 V rms. We will assume that the rated power and power factor are maintained even though the load voltage will be less than 208 V rms. To obtain accurate simulations, we must find the magic value of the load part attribute, Van, such that the source voltage and load power/power factor are satisfied. We can do this in PSPICE very easily.

We will sweep the Van attribute and plot the load complex power. With the cursors, we can find the required Van value. Using that value in a new simulation, we can obtain accurate simulations. The appropriate *Schematics* diagram is shown in Figure 19.3 where the PARAM part creates the variable, Vsweep, and the wye load attribute, Van, has been set to {Vsweep}. The braces inform PSPICE that the value of Van is a function rather than a number. The required Parametric setup is shown in Figure 19.4. After simulating and PROBE opens, we want to plot the load power. Select Add from the Traces menu to open the window in Figure 19.5. Notice the Trace expression at the bottom of the figure. It is the total complex power at the load. The resulting PROBE plot is in Figure 19.6 with the Vsweep value required for operation at 18 kVA – namely, 198.65 V_RMS. Returning to the *Schematics* diagram, we change Van from {Vsweep} to 198.65 V rms and disable the Parametric sweep. From the resulting output file, the line current and line-to-neutral voltage are

$$**** \quad \text{AC ANALYSIS} \quad ****$$

Simulated Quantity	Magnitude (rms)	Phase (degrees)
Line Current	30.21 A	-38.88
Line to Neutral Voltage	198.6 V	-2.010

And, of course, these values yield a total complex power of

$$S_{3\phi} = 3V_{AN}I_{aA} = 3*30.30*198.6 = 18.00 \text{ kVA}$$

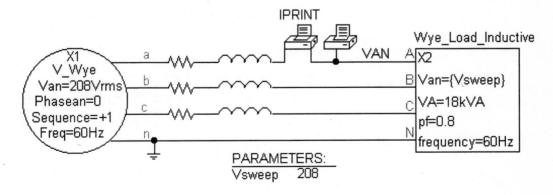

Figure 19.3. The Schematics diagram for simulation 19.3.

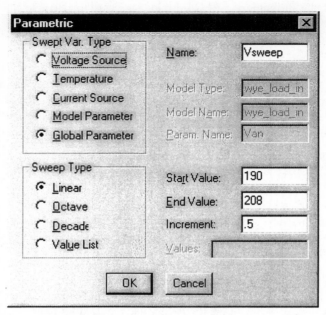

Figure 19.4. The Parametric Setup box edited to vary Vsweep between 190 and 208 V.

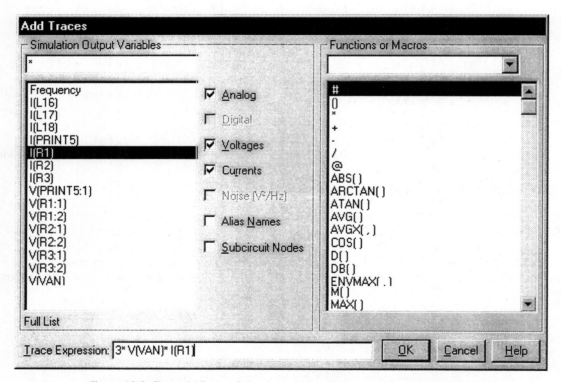

Figure 19.5. The Add Trace dialog box set to plot the total complex power.

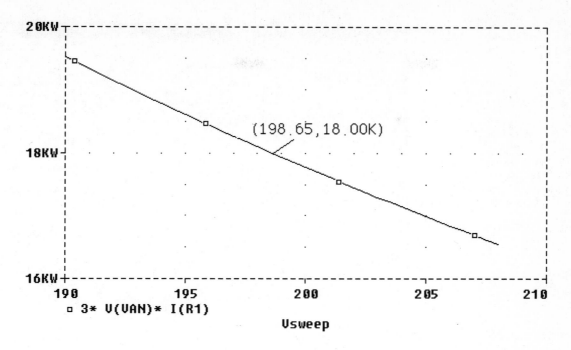

Figure 19.6. PROBE plot showing that when Vsweep = 198.65 V$_{RMS}$, the load complex power is 18 kVA.

19.4 As mentioned in the BECA text, power factor correction is a real concern of the power company. The inductive loads presented by industries (lots of motors) require the utility to generate more energy to supply the same power to those industries. To improve the situation, customers with poor power factors pay a higher $/kW-hour rate. If the rate increase is large enough, the customer will purchase power facto correction capacitors.

In this simulation, we will show how PSPICE can be used to solve power factor correction problems. Conside a wye-wye, 60 Hz, balanced 3-ϕ system with a line-to-neutral voltage of 440 V rms, a line impedance consisting of 5 Ω resistance and 10 mH inductance and an equivalent load of 12 Ω and 100 mH. A power factor between 0.95 and 1.0 lagging is desired. What is the range of the power factor correction capacitor value? The single-phase equivalent *Schematics* diagram is shown in Figure 19.7, where a power factor correction capacitor has been added. Since we wish to plot the power factor versus the capacitor value, the power factor correction capacitor value is set equal to the variable defined in the PARAM part as Cpf. The **Parametric** setup is shown in Figure 19.8 and, of course, a 60- Hz **AC Sweep** analysis is requested. After simulation, to plot the power factor in PROBE, we enter the following **Trace Expression** in the **Add Traces** window.

Trace Expression: cos(P(I(Rline))*3.1416/180)

We should point out that in PROBE signal phases are in degrees but trigonometric function arguments are in radians, which explains the π/180 factor.

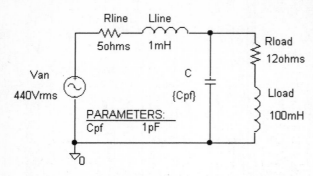

Figure 19.7. *Schematics* diagram for simulation 19.4.

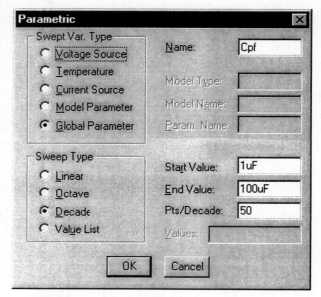

Figure 19.8. Parametric setup to sweep Cpf.

From the simulation results in Figure 19.9, we see that the power factor condition is satisfied for a capacitor value between 57 μF and 63.1 μF. Remember that the simulation is for a single phase. Three capacitors are required for the actual system.

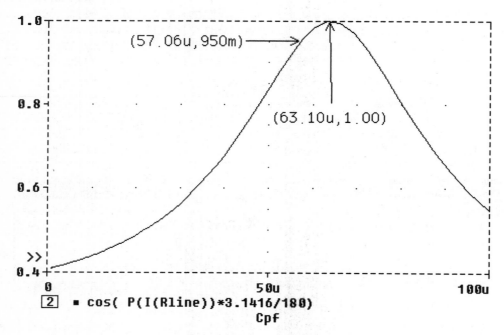

Figure 19.9. Simulation results for simulation 19.4.

20. Simulating Variable Frequency Circuits in PSPICE

Variable frequency network analysis is nothing more than steady-state ac analysis where the frequency of excitation is a variable. The solution contains no transient information and voltages and currents are represented in phasor form. Typically, since frequency is a variable, solutions are shown in Bode plots. We will now demonstrate how to create and utilize Bode plots in both PSPICE and EWB.

The performance of variable frequency circuits, often called filters, is usually depicted in a Bode plot. Bode plots are graphs of the phasor magnitude and phase versus frequency where magnitude is plotted in dB, phase in degrees and frequency on a log axis. It is very easy to create Bode plots in PSPICE, as we will show in the following simulations. However, we will take a moment to introduce the **Markers** menu in *Schematics*. A marker is called a pseudo-part in *Schematics*. While a marker has no effect on the simulation, it does inform PROBE to plot the marker quality. The primary markers are listed in Table 20.1 and shown in Figure 20.1.

Table 20.1

The principle markers available in *Schematics*

Marker	Simulation Type	Plotted Quality
Voltage/Level	dc	node voltage dc value
	ac	node voltage phasor magnitude
	transient	node voltage instantaneous value
Voltage Differential	dc	dc voltage between two nodes
	ac	magnitude of phasor voltage between two nodes
	transient	instantaneous voltage between two nodes
Current into Pin*	dc	dc current into a pin
	ac	current phasor magnitude of current into a pin
	transient	instantaneous current into a pin
Advanced		
vdb	same as Voltage/Level	voltage in decibels
idb*	same as Current into Pin	current in decibels
vphase	ac only	phase of the node voltage
iphase*	ac only	phase of the pin current
*Must be placed directly on the pin of interest.		

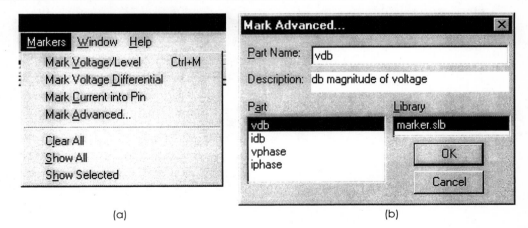

(a)   (b)

Figure 20.1. Accessing the markers in Schematics: (a) the **Markers** menu and (b) some of the advanced markers.

20.1 Let's demonstrate a variable frequency analysis and the use of the markers by simulating the passive low-pass filter in Figure 20.2. We will also produce a Bode plot of the output voltage and extract the bandwidth of the filter and the phase when the frequency equals the bandwidth.

Note that in Figure 20.2, the vdb and vphase advanced markers have been used to create the Bode plot. The frequency will be varied from 1 Hz to 10 kHz in decades (log scale) as shown in the AC Sweep dialog box in Figure 20.3. The resulting Bode plot is shown in Figure 20.4 where the critical data points have been tagged. The bandwidth is found to be about 174 Hz and the phase at 174 Hz is -44.8°.

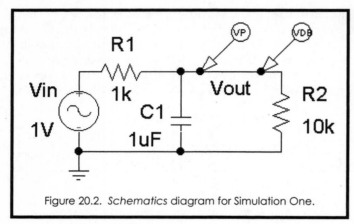

Figure 20.2. *Schematics* diagram for Simulation One.

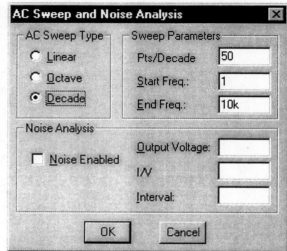

Figure 20.3. The AC Sweep dialog box edited to vary frequency in decades from 1 Hz to 10 kHz.

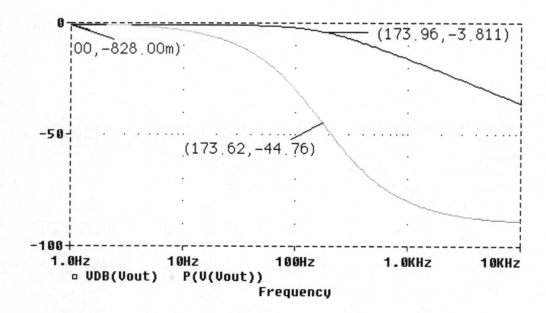

Figure 20.4. The Bode plot for simulation 20.1 showing the data points of interest.

204.2   As a demonstration of resonance, let's simulate the Panasonic ELJ-FC100JF 10-µH inductor that has parasitic resistance and capacitance values of 6.33 Ω and 2.47 pF respectively.

We can find the resonant frequency using the *Schematics* circuit in Figure 20.5. We will vary the frequency of the ac current source until the voltage reaches a maximum. The frequency at which the maximum occurs is the resonant frequency. From the simulation results plotted in Figure 20.6, we find that the resonant frequency of the inductor is 32.36 MHz, which matches the manufacturer's specification of 32 MHz very well. Note that resonance occurs without any other passive elements connected to the inductor. We call this kind of resonance *self-resonance*.

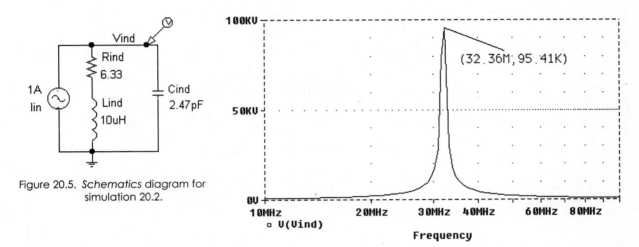

Figure 20.5. *Schematics* diagram for simulation 20.2.

Figure 20.6. PROBE results for simulation 20.2 indicating inductor self-resonance.

20.3 The *Schematics* circuit diagram in Figure 20.7 is a 2nd order low-pass filter. Let's simulate it, using the quasi-ideal opamp in the BECA library, to determine the center frequency, the bandwidth and the Q factor of the filter.

From the resulting PROBE plot, shown in Figure 20.8, we find the center frequency is 501.2 kHz and the 3 dB corner frequencies are 511.8 kHz and 494.8 kHz. Therefore, bandwidth and Q factor are

$$BW = 511.8k - 494.8k = 7.0 \text{ kHz}$$

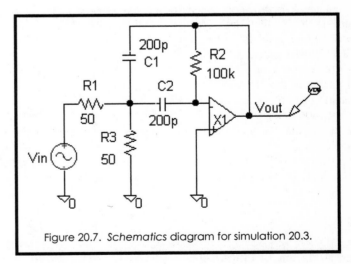

Figure 20.7. *Schematics* diagram for simulation 20.3.

$$Q = \frac{f_O}{BW} = \frac{501.2k}{7.0k} = 71.6$$

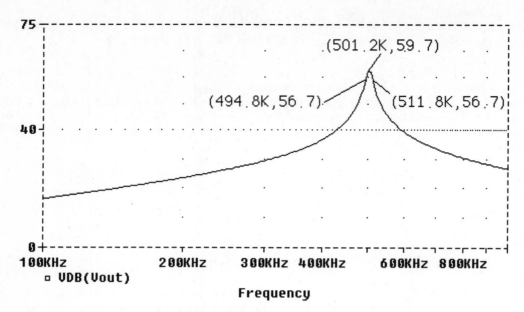

Figure 20.8. PROBE results for simulation 20.3.

20.4 For the final PSPICE simulation of this section, we will do a Performance Analysis simulation. In this case, we will simulate the 2nd order bandpass filter in Figure 20.9 to determine the value of R_B that yields the highest Q value.

This requires an AC Sweep across a range of frequencies, a Parametric sweep of R_B and a means of plotting Q. Figure 20.10 shows the corresponding *Schematics* diagram. Using the PARAM part and setting the AC Sweep and the Parametric dialog boxes as shown in Figure 20.11 satisfies the first two requirements. After simulating and opening PROBE, select Performance Analysis from the Trace menu. This causes the variable R to become the *x*-axis variable.

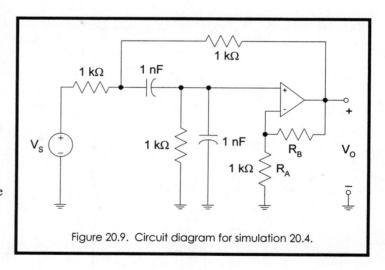

Figure 20.9. Circuit diagram for simulation 20.4.

Now we will plot the quality factor, Q, which we know to be

$$Q = \frac{center\ frequency}{bandwidth}$$

Select Add from the Trace menu to open the window in Figure 20.12. On the right side of the window is a list of goal functions. We are interested in two in particular: CenterFreq(1,db_level) and BPBW(1,db_level). Let's examine the arguments of these goal functions. First, for CenterFreq(1,db_level), the number 1 indicates that CenterFreq is a function of only one voltage or current. The remaining variable, db_level, is just a number – we recommend 0.1. To demonstrate how the CenterFreq(1,db_level) goal function works, consider the following example

CenterFreq(V(Vout),0.1)

90

PROBE finds the maximum value of the voltage Vout. Next, the frequencies that correspond to that maximum minus 0.1 dB are found. The center frequency is then calculated as the average of these two frequencies. The accuracy of this averaging approach improves as the **db_level** value is decreased.

Arguments for the **BPBW(1,db_level)** function (bandpass bandwidth) are exactly the same. Since bandwidth is specified at 3 dB below the maximum value, we should set db_level = 3. Plots of center frequency and bandwidth are shown in Figure 20.13a. Notice that the center frequency is fairly

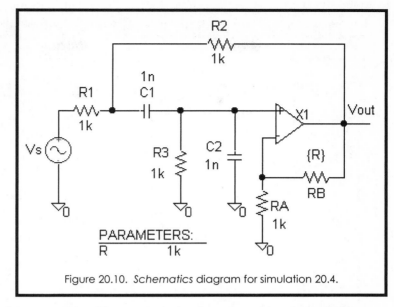

Figure 20.10. *Schematics* diagram for simulation 20.4.

constant while the bandwidth depends on R_B. This will cause Q to also depend on R_B as shown in Figure 20.13b. We find that Q has a maximum value of 21.8 when $R_B = 4$ kΩ.

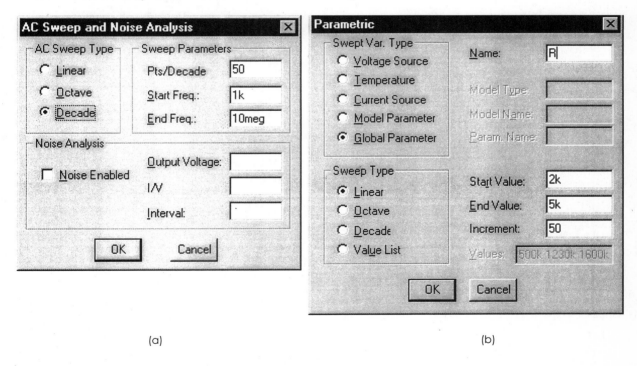

(a)   (b)

Figure 20 11. Setup dialog boxes edited for simulation 20.4: (a) AC Sweep and (b) Parametric.

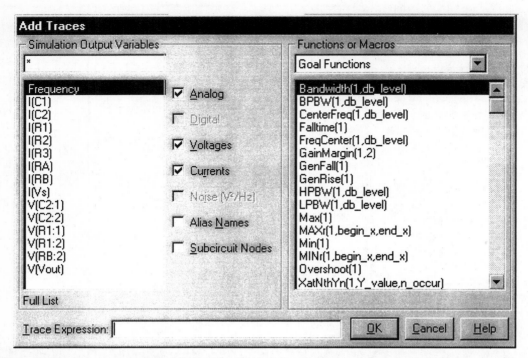

Figure 20.12. The Performance Analysis Add Traces window. Goal functions are on the right while voltages and currents are on the left.

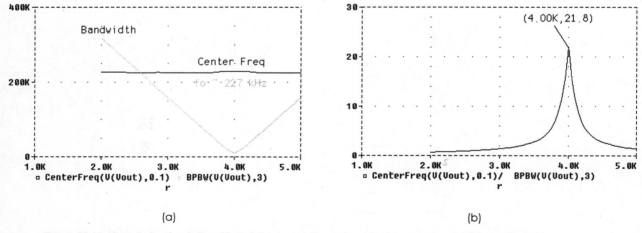

(a)   (b)

Figure 20.13. Results for simulation 20.4: (a) bandwidth and center frequency and (b) quality factor.

21. MATLAB's Laplace Transform Operation

We will now demonstrate the simplicity of using MATLAB to (a) find the Laplace transform of a given time function and (b) compute the inverse Laplace transform.

21.1 Using MATLAB's `laplace` operation, let us find the Laplace transform of the following functions.

a) $f(t) = 14e^{-5t} u(t)$

b) $f(t) = \left(2e^{-4t} - 6e^{-t} + 14e^{-3t}\right)u(t)$

c) $f(t) = 17e^{-500t} sin(10^5 t) \, u(t)$

d) $f(t) = 100 sin^2(25t) \, u(t)$

a) In MATLAB, we first define the symbols we will be using

```
» syms t
```

Now we enter the functions listed above.

```
» laplace(14*exp(-5*t))

    ans =
         14/(s+5)
```

The answer is $\mathbf{F}(s) = \dfrac{14}{s+5}$

b)

```
» laplace(2*exp(-4*t)-6*exp(-t)+14*exp(-3*t))

    ans =
    2/(s+4)-6/(s+1)+14/(s+3)
```

The answer is $\mathbf{F}(s) = \dfrac{2}{s+4} - \dfrac{6}{s+1} + \dfrac{14}{s+3}$

c)

```
» laplace(17*exp(-500*t)*sin(10^5*t))

    ans =
    1700000/((s+500)^2+10000000000)
```

The answer is $\mathbf{F}(s) = 17\dfrac{10^5}{(s+500)^2 + (10^5)^2}$

d)

```
» laplace(100*(sin(25*t))^2)

    ans =
    125000/s/(s^2+2500)
```

The answer is $\mathbf{F}(s) = \dfrac{125000}{s(s^2+2500)}$

21.2 Using MATLAB's `ilaplace` operation, find the inverse Laplace transform of the following functions.

a) $F(s) = \dfrac{s}{s^2 + 10^6}$

b) $F(s) = \dfrac{s + 0.1}{s^2 + 0.2s + 100.01}$

c) $F(s) = \dfrac{s + 5}{(s+1)^2(s+3)}$

d) $F(s) = \dfrac{s^3 + 5s^2 - 50s - 100}{s(s^2 + 15s + 50)}$

a)   In MATLAB, we first define the symbols we will be using

```
» syms s
```

Now we enter the functions listed above.

```
» ilaplace(s/(s^2+10^6))

    ans =
          cos(1000000^(1/2)*t)
```

The answer is $f(t) = \cos(1000t)\,u(t)$.

b)

```
» ilaplace((s+0.1)/(s^2+0.2*s+100.01))

    ans =
          exp(-1/10*t)*cos(10*t)
```

The answer is $f(t) = e^{-t/10}\cos(10t)\,u(t)$.

c)

```
» ilaplace((s+5)/((s+1)^2*(s+3)))

    ans =
           2*t*exp(-t)-1/2*exp(-t)+1/2*exp(-3*t)
```

The answer is $f(t) = 2te^{-t} - 0.5e^{-t} + 0.5e^{-3t}\,u(t)$.

d)

```
» ilaplace((s^3+5*s^2-50*s-100)/((s^2+15*s+50)*s))

    ans =
           Dirac(t)-2*exp(-10*t)-6*exp(-5*t)-2
```

The answer is $f(t) = [\delta(t) - 2e^{-10t} - 6e^{-5t} - 2]\,u(t)$.

22. Applications of the Laplace Transform

Using the Laplace transform in circuit analysis is the culmination of our work. It converts differential equations to algebraic equations. Of course, the frequency domain does the same thing. But the frequency domain is limited to steady-state ac systems while the s-domain is applicable to dc, transient and ac analyzes. Also, as you become more comfortable with the s-domain, you will find that the mathematics readily surrenders information about the circuit under inspection.

23. PSPICE Simulations in the s-domain

23.1 For our first simulation, we will demonstrate the generic relationship between the time and s-domains for systems described by simple differential equations.

A time-honored application for electronic circuitry, both digital and analog, has been calculating the trajectories of artillery fired projectiles. In fact, ENIAC, the world's first true digital computer, was constructed for the sole purpose of calculating such trajectories. We will instead use an analog approach to plot range versus altitude and range versus time for projectiles fired from the artillery piece depicted in Figure 23.1. We assume that at $t = 0$, the projectile is launched with particular horizontal and vertical

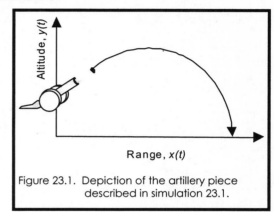

Figure 23.1. Depiction of the artillery piece described in simulation 23.1.

velocities. Also, the projectile has no means of self-propulsion. From our freshman physics classes, we know that the range $x(t)$ and altitude $y(t)$ are given by

$$x(t) = \int \dot{x}(0)\, dt \qquad\qquad y(t) = y(0) + \int \dot{y}(0)\, dt - \int\int g\, dt^2 \qquad\qquad (23.1)$$

where g is the gravitational constant and the dotted terms represent velocities in the x and y directions.

To develop a circuit that can model the system described in (23.1), we will make use of the inverting integrator shown in Figure 23.2. Remember that for a particular simulation, the initial velocities, altitude and g are constants. The circuit in Figure 23.3 models the equations in (23.1) very well. Each op-amp subcircuit is either an integrator, a summer or a summing integrator. You should be able to verify that,

$$v_X(t) = \int v_{VXO}\, dt \qquad v_Y(t) = v_{YO} + \int v_{VYO}\, dt - \int\int v_G\, dt^2 \qquad (23.2)$$

Comparing (23.1) to (23.2), we see the following analogies

Equation 23.1	Equation 23.2
$x(t)$	$v_X(t)$
$\dot{x}(0)$	v_{VXO}
$y(t)$	$v_Y(t)$
$y(0)$	v_{YO}
$\dot{y}(0)$	v_{VYO}
g	$v_G(t)$

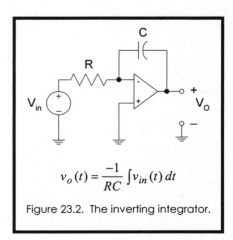

$$v_o(t) = \frac{-1}{RC} \int v_{in}(t)\, dt$$

Figure 23.2. The inverting integrator.

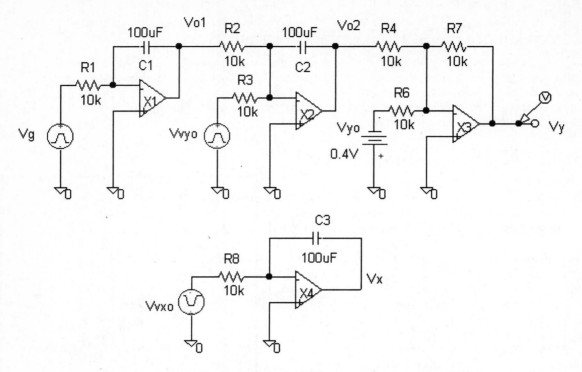

Figure 23.3. Schematics diagram for the analog computer in simulation 23.1.

To complete the analogy, we must decide how many kilometers on the firing range correspond to one volt in the circuit. Here, we have chosen to let one volt represent one km. Although it is an option, we have decided not to scale the simulation time axis.

Now for a particular simulation. The projectile is launched at a 45° angle with an initial velocity of 500 m/s. Also, the artillery piece itself is at 400 m above ground level. We will find the range versus time and the trajectory. Since $g = 9.8$ m/s^2, and 1 volt corresponds to 1 km, we find $V_G = 9.8$ mV. Also, due to the 45° launch angle, the initial x and y velocities are equal, and

$$\dot{x}(0) = \dot{y}(0) = 500\cos(45) = 353.5 \text{ m/s} \qquad v_{VXO} = v_{VYO} = 0.3535 \text{ V}$$

Similarly, an initial altitude of 400 m corresponds to $v_{yo} = 0.4$ V.

The simulation results for range versus time are shown in Figure 23.4. Using the cursor, the maximum range and time-to-target are 25.9 km and 73.2 seconds. Plotting the trajectory is a two step process. First, delete Vx and add the trace, Vy. Second, change the x-axis variable from time to Vx. Do this by selecting X Axis Settings from the Plot menu in PROBE. The dialog box in Figure 23.5 opens. Select Axis Variable. When the X Axis Variable dialog box appears, simply choose V(Vx). The resulting trajectory plot is shown in Figure 23.6, where we find the maximum altitude is 6.77 km.

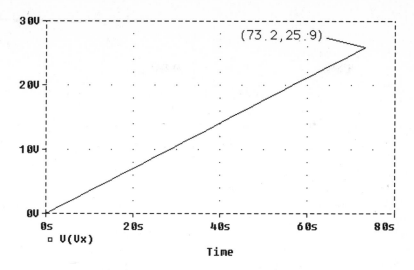

Figure 23.4. PROBE results for range and time-to-target for the artillery piece in simulation 23.1.

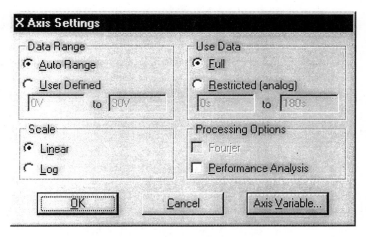

Figure 23.5. The X Axis Settings dialog box.

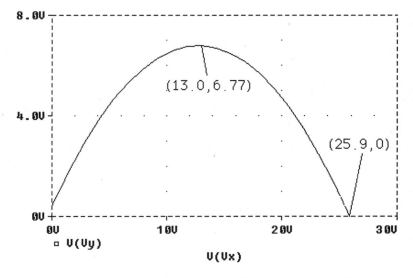

Figure 23.6. Projectile trajectory for the artillery piece in simulation 23.1.

23.2 In this simulation, we will examine the relationship between the Bode plot and the *s*-plane for the 2nd order network in Figure 23.7. In particular, we are interested in the pole locations of the transfer function for over-, under- and critical damping.

By varying the resistor, we can produce each of these cases. Here, R has been stepped from 0.5 Ω to 5 Ω in 1.5 Ω increments. Critical damping occurs at 2 Ω. Figure 23.8 shows the resulting Bode plots (magnitude only) which were created in PSPICE using the PROBE utility. The conspicuous "hump" in the Bode plot makes the underdamped case stand out. As for the critically damped and the two overdamped cases, they are difficult to distinguish because the poles are close to one another. When R = 5 Ω, the poles are farthest apart at 4.79 and 0.21 rad/s. But this is not as obvious on a Bode plot.

The *s*-plane plots for R = 0.5 Ω, 2 Ω and 5 Ω are shown in Figure 23.9. When the network is underdamped, R = 0.5 Ω, we see the poles are complex conjugates of one another, as expected. For critical damping, the poles are real and equal. And, as the damping increases and the system is more and more overdamped, the poles are real and both move farther and farther apart.

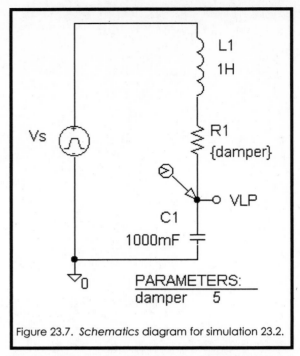

Figure 23.7. *Schematics* diagram for simulation 23.2.

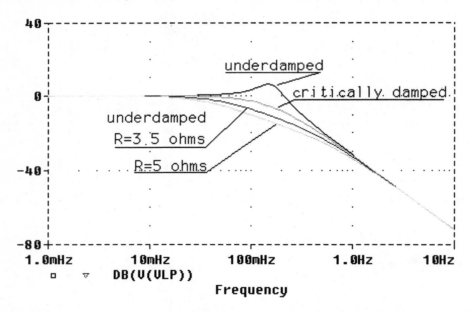

Figure 23.8. Effect of damping on the Bode plot (magnitude only) for the network in simulation 23.2.

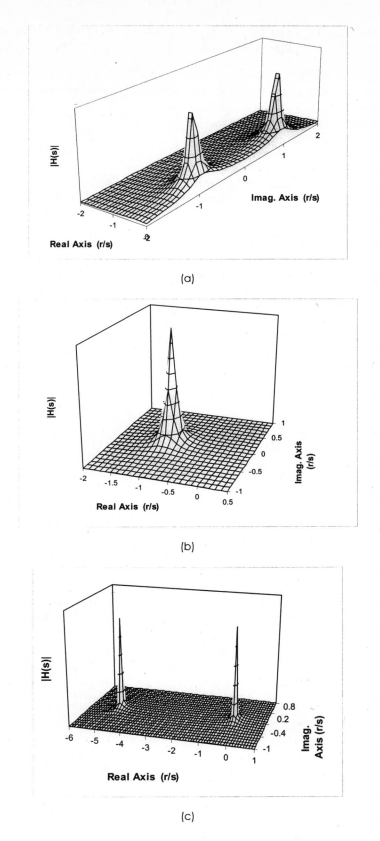

Figure 23.9. *s*-plane plots for (a) R = 0.5 Ω, (b) R = 2 Ω and (c) R = 5 Ω.

23.3 Sometimes we know a circuit's transfer function in the s-domain rather than its topology (components and connections). For simulation purposes, we could develop an equivalent network that has the same transfer function, or we can use the LAPLACE part in the ABM library.

The LAPLACE part, shown in Figure 23.10, has an input, an output and two attributes: the transfer function numerator and denominator. Both the numerator and denominator can be functions of *s* but they cannot be functions of time or any other voltage or current. As an example, the part in Figure 23.10 models a 1st order lowpass filter with a pole frequency of 1.59 kHz (*s* is in rad./s) and a dc gain of 100 (40 dB). The Bode plot in Figure 23.11 verifies the model.

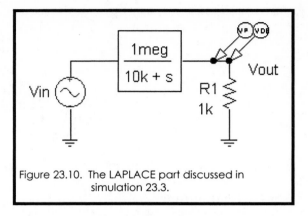

Figure 23.10. The LAPLACE part discussed in simulation 23.3.

The LAPLACE part is modeled differently in each of the three PSPICE analysis types – dc, ac and transient. In a dc analysis, the output is the dc input times transfer function evaluated at *s* = 0. For an ac analysis, the variable *s* is set equal to jω and the transfer function becomes just another complex number. Finally, in a transient analysis, the output signal is the convolution of the input and the part's impulse response. Ouch! A forewarning, performing these calculations can be rather time consuming in a network with multiple LAPLACE parts.

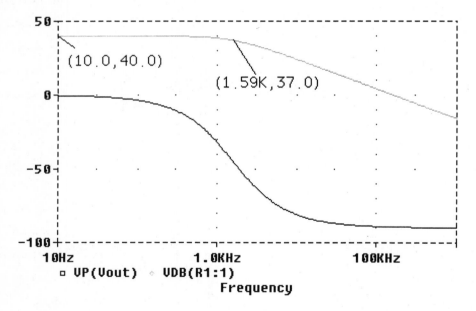

Figure 23.11. Bode plot for the lowpass filter modeled by the LAPLACE part in simulation 23.3.

24. The Fourier Series and PSPICE

PSPICE can determine the Fourier series of any node voltage or branch current. A transient simulation must be performed since the Fourier series is calculated from time-domain data points. The series data, consisting of the dc component, the fundamental and harmonic magnitudes and their phases, is tabulated in the output file. Note that PSPICE uses a trigonometric series of sinewaves, not cosinewaves.

24.1 To demonstrate this feature, we will find the Fourier series for the simple circuit in Figure 24.1 where

$$v_{1k}(t) = 2\sin(2\pi \cdot 1000t) \text{ V} \qquad v_{3k}(t) = 0.5\sin(2\pi \cdot 3000t) \text{ V} \qquad V_{DC} = 5 \text{ V} \qquad (24.1)$$

To request a Fourier series, go to the **Transient** dialog box and select **Enable Fourier**, as shown in Figure 24.2. As for the editable fields, **Center Frequency** is the same as fundamental frequency. It should be the same as the lowest frequency in the waveform of interest, 1 kHz in this example. **Number of harmonics** is exactly that – how many harmonics are calculated and listed in the output file. **Output Vars** is a listing of the node voltages and branch currents you wish to include in the Fourier analysis. Examples of suitable node voltage variables are V(3), V(Vo) and V1(R1). Entries such as Vo will not work. Currents should be entered as I(R1), I(V1), etc.

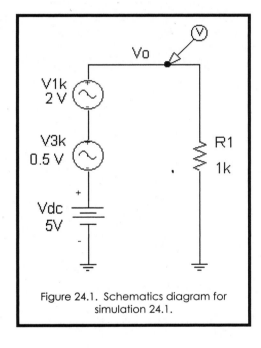

Figure 24.1. Schematics diagram for simulation 24.1.

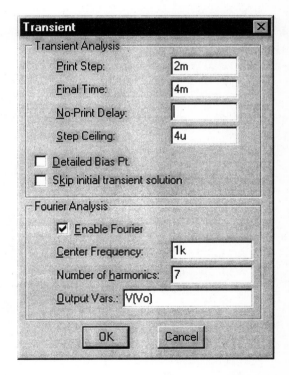

Figure 24.2. The Transient dialog box edited to find the first 7 Fourier components.

How does PSPICE determine the series? Your **Center Frequency** entry is inverted to produce the fundamental period, $T_{FUND} = 1/f_C$. The data points from the last T_{FUND} time span are extracted from the PROBE data file. Assuming the extracted data to be periodic, PSPICE constructs the series. This means that the duration of your transient simulations should extend at least one fundamental period beyond the point where transient behavior has died out. In general, increasing the number of data points by reducing the **Step Ceiling** in Figure 24.2 will improve the accuracy of the calculations.

Back to our simulation. The PROBE plot for Vo is shown in Figure 24.3. From the output file, the Fourier series data is listed below.

```
--------------------------------------------------------------------------------------
FOURIER COMPONENTS OF TRANSIENT RESPONSE V(Vo)

  DC COMPONENT =    5.000000E+00

  HARMONIC    FREQUENCY       FOURIER      NORMALIZED        PHASE       NORMALIZED
    NO          (HZ)        COMPONENT     COMPONENT         (DEG)      PHASE  (DEG)

     1        1.000E+03     2.000E+00     1.000E+00      1.167E-05      0.000E+00
     2        2.000E+03     1.797E-05     8.987E-06      3.711E+01      3.711E+01
     3        3.000E+03     4.998E-01     2.499E-01      4.214E-05      3.047E-05
     4        4.000E+03     8.018E-06     4.009E-06      3.945E+01      3.945E+01
     5        5.000E+03     6.420E-07     3.210E-07      1.445E+02      1.445E+02
     6        6.000E+03     8.032E-06     4.016E-06     -1.475E+02     -1.475E+02
     7        7.000E+03     4.242E-06     2.121E-06     -1.124E+02     -1.124E+02


         TOTAL HARMONIC DISTORTION =    2.498946E+01 PERCENT
--------------------------------------------------------------------------------------
```

From the series listing, we see that the dc component, the first harmonic (same as the fundamental) and the third harmonic match the equations in (24.1). Although the other harmonics are non-zero, they are so extremely small that they have negligible effect on the series.   The harmonic distortion is calculated as follows

$$harmonic\ distortion = \frac{\sqrt{v_2^2 + v_3^2 + v_4^2 + \cdots + v_N^2}}{v_1} \tag{24.2}$$

where v_i is the magnitude of the ith harmonic.

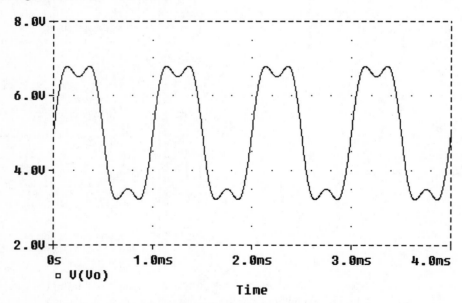

Figure 24.3. PROBE transient for Vo in simulation 24.1.

102

24.2 A European phone line is picking up 50 Hz noise from the power grid. Let us design a filter to remove the noise and verify its performance in PSPICE.

The notch filter in Figure 24.4 is designed to extract the noise without altering frequencies above 250 Hz. The voltage sources Vnoise and Vin model the 50 Hz noise and the phone conversation respectively and are given by the expressions

$$v_{noise} = 1\sin(2\pi \cdot 50t) \text{ V}$$

$$(24.3)$$

$$v_{in} = 0.2\sin(2\pi \cdot 250t) \text{ V}$$

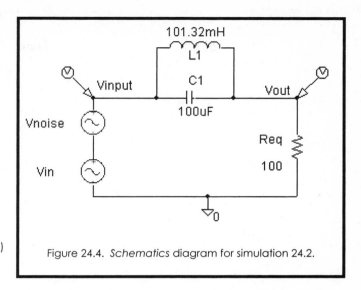

Figure 24.4. *Schematics* diagram for simulation 24.2.

The key to the notch filter is the LC parallel combination. At the resonance frequency, their impedances are equal in magnitude but opposite in sign. Performing the standard parallel combination calculation yields an infinite impedance at resonance. As a result, no energy is transferred to the output.

We will use three simulation results to verify the performance of the filter: a Bode plot of the filter transfer function, a time-domain (transient) plot of the input and output, and an inspection of the Fourier series for the input and output voltages (center frequency = 50 Hz). Figure 24.5 shows the Bode plot of the filter demonstrating that the notch indeed occurs at 50 Hz and signals at 250 Hz are essentially unaltered. Transient simulation results are shown in Figure 24.6. After about 70 ms, all transients have died out, and the output is a 250 Hz sinewave – no 50 Hz noise is present. From the Fourier series data, that follows, we see that the 50 Hz component of the output voltage is roughly 1% of the input voltage component. Also, the 250 Hz component (5th harmonic) is within 1% of the input value. Looks like our filter is a winner.

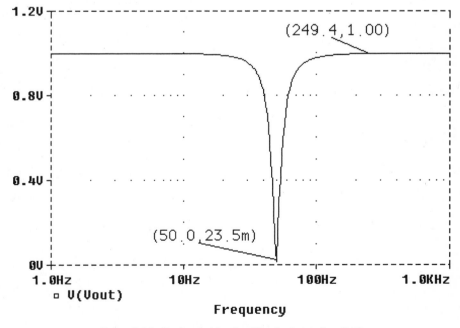

Figure 24.5. Bode plot for the filter in simulation 24.2.

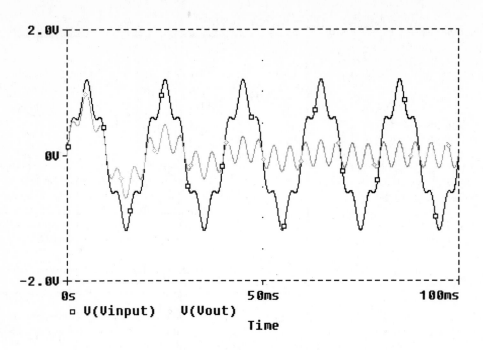

Figure 24.6. Transient results for the filter in simulation 24.2.

```
------------------------------------------------------------------------------------------
FOURIER COMPONENTS OF TRANSIENT RESPONSE V(Vout)

DC COMPONENT =    1.692078E-03

HARMONIC   FREQUENCY    FOURIER     NORMALIZED      PHASE      NORMALIZED
   NO        (HZ)      COMPONENT    COMPONENT       (DEG)      PHASE (DEG)

    1       5.000E+01   1.113E-02   1.000E+00     -1.721E+01    0.000E+00
    2       1.000E+02   1.303E-03   1.171E-01     -1.101E+02   -9.293E+01
    3       1.500E+02   5.779E-04   5.191E-02     -1.265E+02   -1.093E+02
    4       2.000E+02   3.613E-04   3.246E-02     -1.366E+02   -1.194E+02
    5       2.500E+02   1.993E-01   1.790E+01      3.751E+00    2.096E+01

    TOTAL HARMONIC DISTORTION =    1.790308E+03 PERCENT


------------------------------------------------------------------------------------------
FOURIER COMPONENTS OF TRANSIENT RESPONSE V(Vinput)

DC COMPONENT =   -2.916756E-07

HARMONIC   FREQUENCY    FOURIER     NORMALIZED      PHASE      NORMALIZED
   NO        (HZ)      COMPONENT    COMPONENT       (DEG)      PHASE (DEG)

    1       5.000E+01   1.000E+00   1.000E+00     -2.014E-04    0.000E+00
    2       1.000E+02   5.430E-07   5.430E-07      3.242E+01    3.242E+01
    3       1.500E+02   6.536E-07   6.536E-07     -7.740E+01   -7.740E+01
    4       2.000E+02   5.071E-07   5.071E-07      1.741E+02    1.741E+02
    5       2.500E+02   1.999E-01   1.999E-01     -8.944E-04   -6.930E-04

    TOTAL HARMONIC DISTORTION =    1.999369E+01 PERCENT
```

25. The Fourier Transform and PSPICE

Sometimes we prefer to view the Fourier characteristics of a network in graphical form rather than tabular. We want a Fourier transform rather than a series. In such cases, use the Fast Fourier Transform in the PROBE utility.

The Fast Fourier Transform (FFT) is activated by clicking on the FFT hotbutton. ![FFT]

Before we look at an FFT example, we must discuss the differences in the FFT and Fourier Series algorithms. First, as mentioned earlier, a Fourier Series is calculated using the last fundamental period of data. In the FFT, all data is used and is assumed to be periodic. Second, the duration of the transient simulation has no effect on the Fourier series accuracy. In the FFT, accuracy and resolution in particular are directly related to the simulation **Final Time**. Also, choosing a simulation **Final Time** that is an integer number of waveform periods will improve the accuracy of the results. Finally, the FFT feature can be used on an expression such as V(Vo)*I(R1).

25.1 Let's use the circuit in Figure 24.1 to investigate the effect of simulation duration on FFT resolution.

From simulation 24.1, we know that the output voltage has components at dc, 1 kHz and 3 kHz with magnitudes of 5V, 1V and 0.5V respectively. We will perform transient simulations of 2 ms (2 fundamental periods) and 200 ms duration and compare the FFT results. In both simulations, the **Step Ceiling** was set equal to the **Final Time**/100. This ensures that both simulations have the same number of data points. Figure 25.1 shows the resulting FFTs. Obviously, the longer simulation has the better resolution. Of course, as the **Final Time** increases, simulations take longer to run. The 2 ms simulation process can be viewed in the Visual Tutor FFT.EXE.

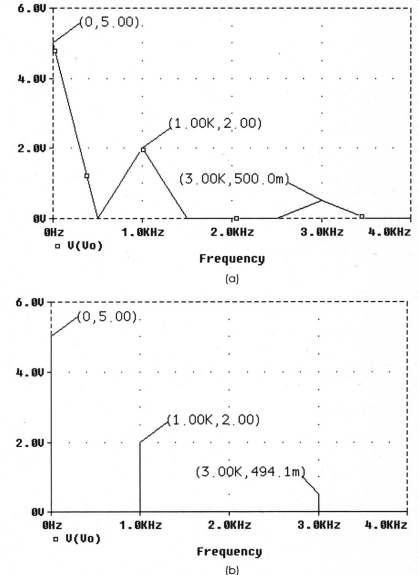

Figure 25.1. FFT results for Final Times of (a) 2 ms and (b) 200 ms.

26. Problem-Solving Videos

The solutions to the problems in this chapter can be viewed within the Circuit Solutions website. Please go to http://www.justask4u.com/irwin and use the Circuit Solutions powered by JustAsk registration code included in the front of your textbook.

CHAPTER 1

P1.24. Find the power that is absorbed or supplied by the network elements.

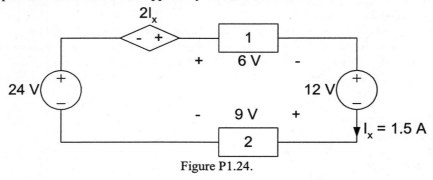

Figure P1.24.

P1.27. Find I_x.

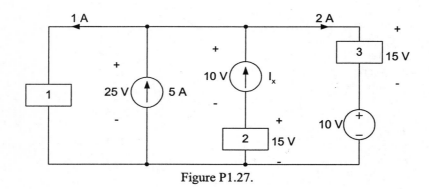

Figure P1.27.

CHAPTER 2

P2.6. The power absorbed by G_x is 50 mW. Find G_x.

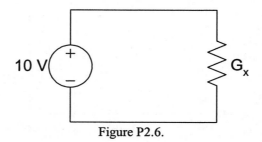

Figure P2.6.

P2.14. Find I_x and I_1.

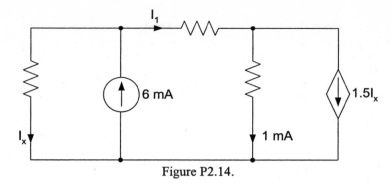

Figure P2.14.

P2.22. Find V_A.

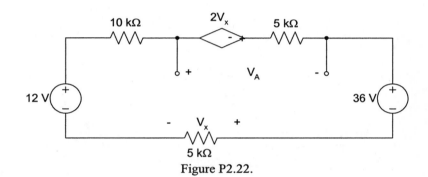

Figure P2.22.

P2.26. Find V_1.

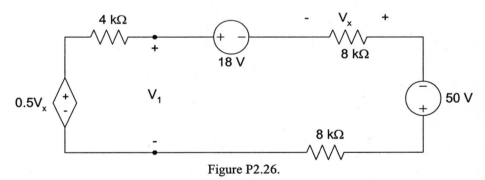

Figure P2.26.

P2.40. Find I_0.

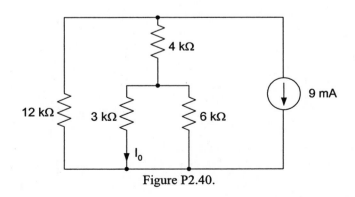

Figure P2.40.

P2.46. Find I_x.

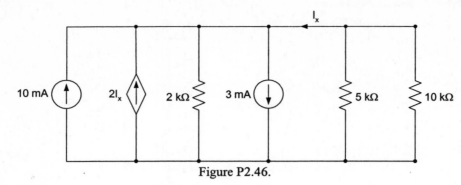

Figure P2.46.

P2.50. Find R_{AB}.

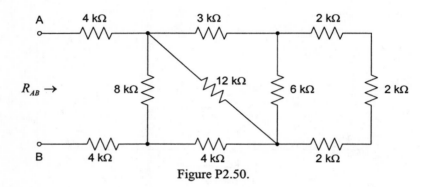

Figure P2.50.

P2.65. Find V_1.

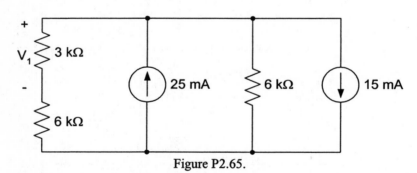

Figure P2.65.

P2.69. If the power supplied by the 3-A current source is 12 W, find V_s and the power supplied by the 10-V source.

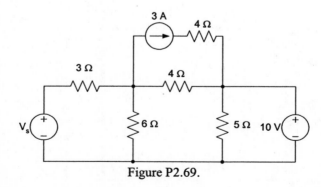

Figure P2.69.

P2.76. Find V_0, V_1, and V_2.

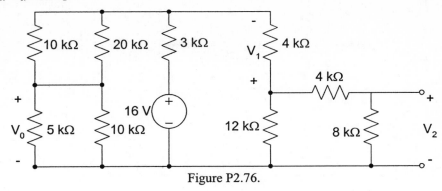

Figure P2.76.

P2.82. Find V_0 and V_1.

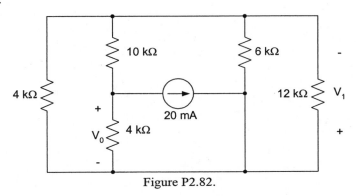

Figure P2.82.

P2.105. Find I_1.

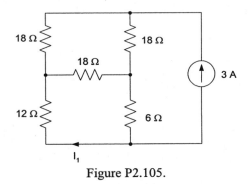

Figure P2.105.

P2.112. Find V_0.

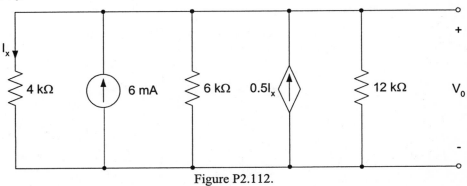

Figure P2.112.

P3.3. Use nodal analysis to find V_0.

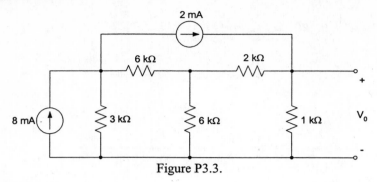

Figure P3.3.

P3.8. Find V_0 using nodal analysis.

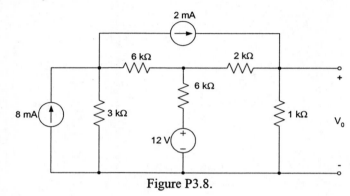

Figure P3.8.

P3.28. Find V_0 using nodal analysis.

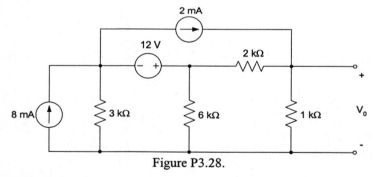

Figure P3.28.

P3.47. Find V_0 using nodal analysis.

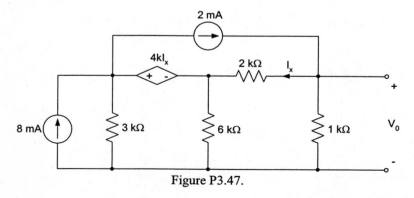

Figure P3.47.

P3.50. Find V_0 using nodal analysis.

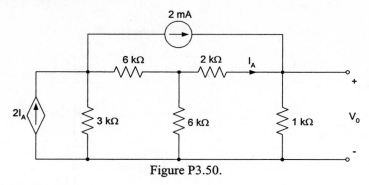

Figure P3.50.

P3.63. Find V_0 using mesh analysis.

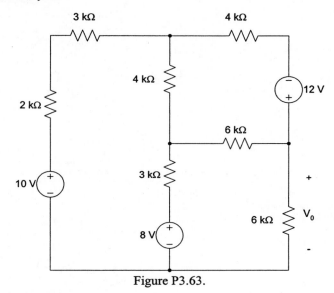

Figure P3.63.

P3.68. Find V_0 using loop analysis.

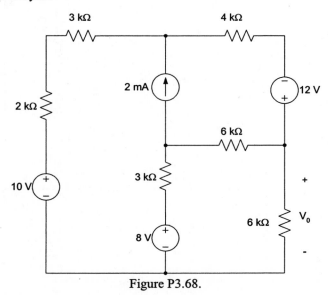

Figure P3.68.

P3.72. Find V_0 in Figure P3.68 using mesh analysis.

P3.83. Find V_0 using mesh analysis.

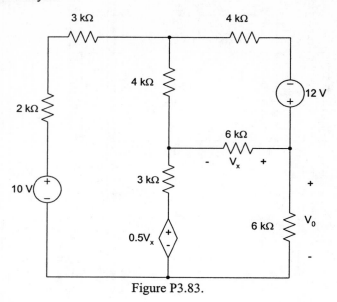

Figure P3.83.

P3.89. Find V_0 using mesh analysis.

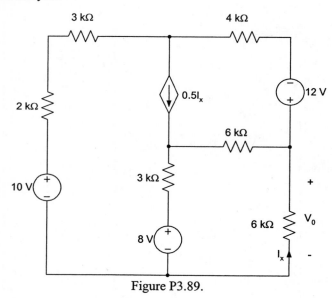

Figure P3.89.

CHAPTER 4

P4.12. Find I_1, I_2, I_3, and I_4.

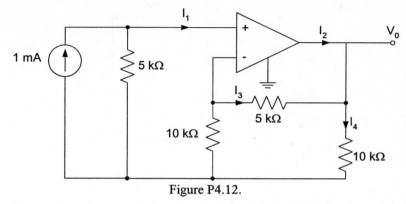

Figure P4.12.

P4.17. a) Find V_0 in terms of V_1 and V_2. b) If $V_1 = V_2 = 4$ V, find V_0. c) If the op-amp power supplies are ± 15 V and $V_2 = 2$ V, what is the allowable range of V_1?

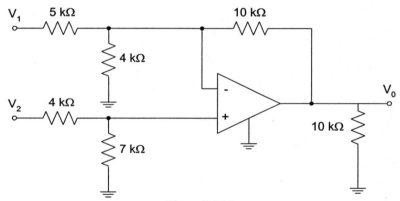

Figure P4.17.

P4.35. Find V_0 and V_3.

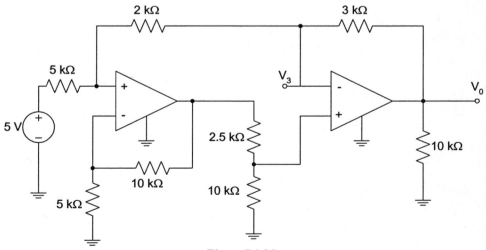

Figure P4.35.

113

P4.38. Find v_0.

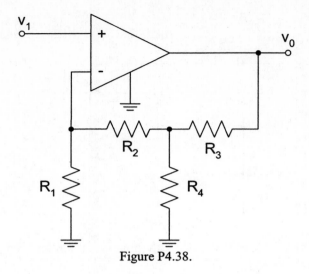

Figure P4.38.

CHAPTER 5

P5.3. Find V_0 using linearity and the assumption that $V_0 = 1$ V.

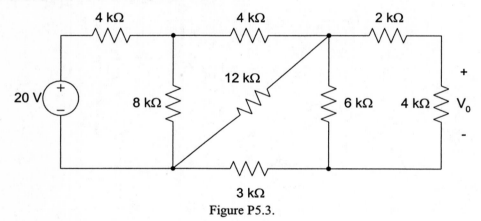

Figure P5.3.

P5.8 Find V_0 using superposition.

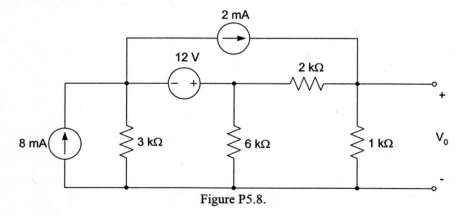

Figure P5.8.

P5.22. Find I_0 using superposition.

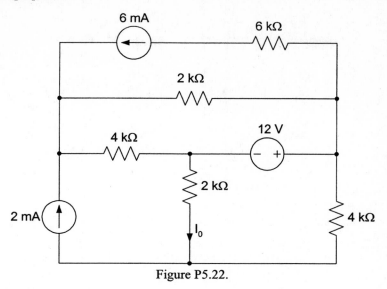

Figure P5.22.

P5.31. Find V_0 using Thevenin's Theorem.

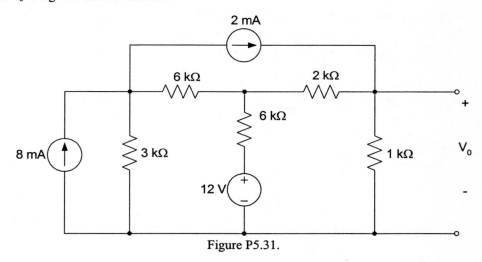

Figure P5.31.

115

P5.58. Find I_0 using Thevenin's Theorem.

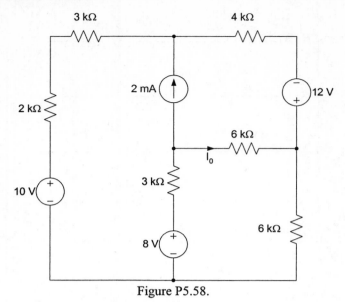

Figure P5.58.

P5.60. Find the Thevenin equivalent of the network at terminals A-B.

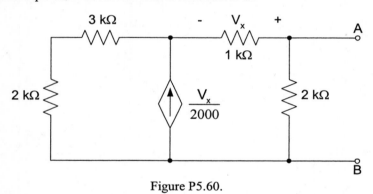

Figure P5.60.

P5.63. Find V_0 using Thevenin's Theorem.

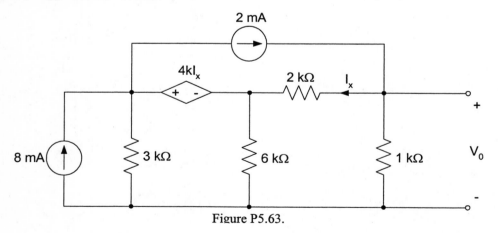

Figure P5.63.

P5.76. Use Thevenin's Theorem to find the power supplied by the 12-V source.

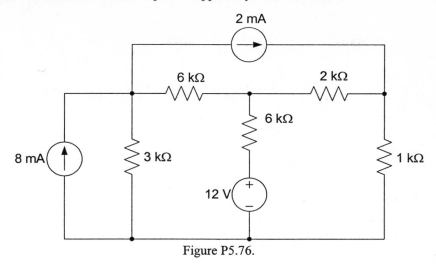

Figure P5.76.

P5.80. Find I_0 using source transformations.

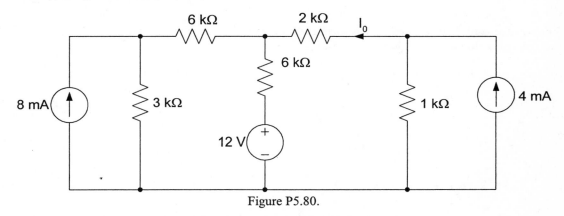

Figure P5.80.

P5.96. Find R_L for maximum transfer and the maximum power transferred to R_L.

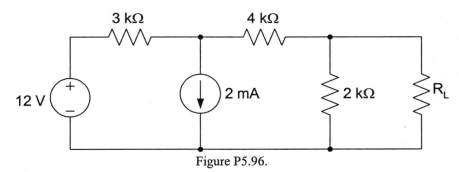

Figure P5.96.

P5.99. Find R_L for maximum transfer and the maximum power transferred to R_L.

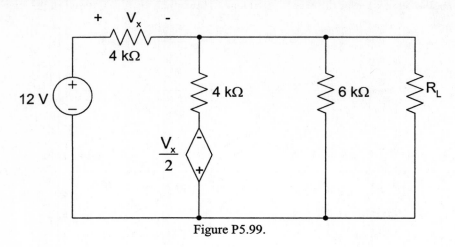

Figure P5.99.

CHAPTER 6

P6.11. The voltage across a 5-μF capacitor is shown below. Find the waveform for the current in the capacitor. How much energy is stored in the capacitor at t = 4 ms?

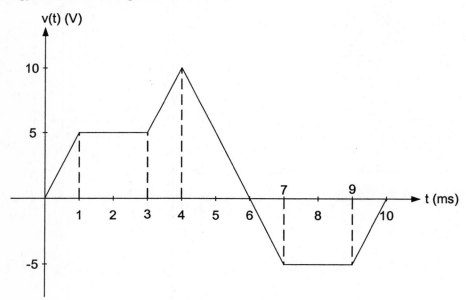

Figure P6.11.

P6.17. The waveform for the current in a 1-nF capacitor is shown below. If the capacitor has an initial voltage of -5 V, determine the waveform for the capacitor voltage. How much energy is stored in the capacitor at t = 6 ms?

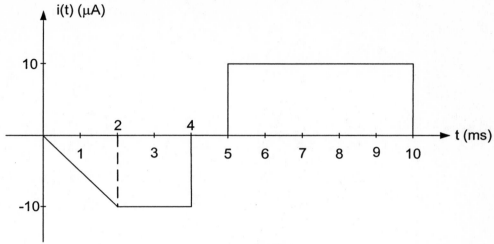

Figure P6.17.

P6.28. The current in a 2-H inductor is shown below. Find the waveform for the inductor voltage. How much energy is stored in the inductor at t = 3 ms?

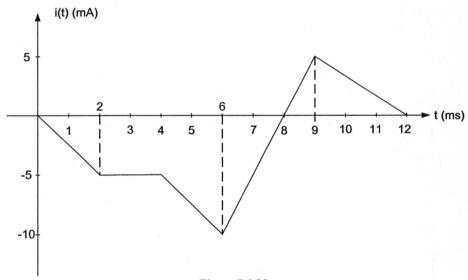

Figure P6.28.

119

P6.34. The voltage across a 0.1-H inductor is shown below. Compute the waveform for the current in the inductor if i(0) = 0.1 A. How much energy is stored in the inductor at t = 7 ms?

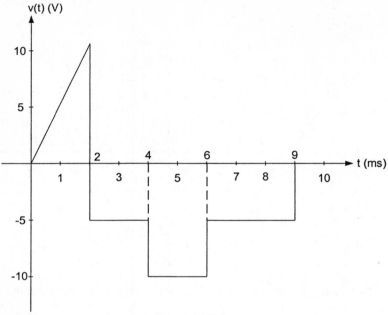

Figure P6.34.

P6.43. Find the energy stored in the capacitor and inductor.

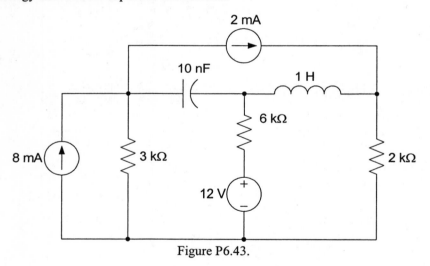

Figure P6.43.

P6.53. Determine C_T in the network below.

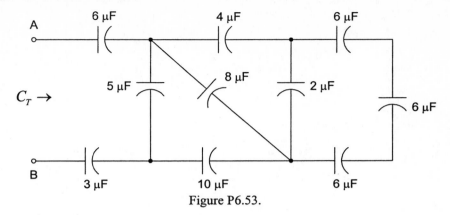

Figure P6.53.

P6.66. Find L_T in the network below.

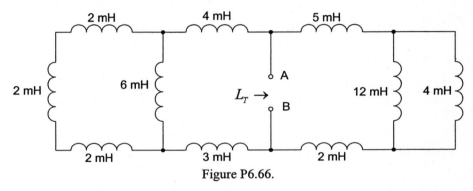

Figure P6.66.

CHAPTER 7

P7.14. Use the differential equation approach to find $v_0(t)$ for $t > 0$. Plot the response.

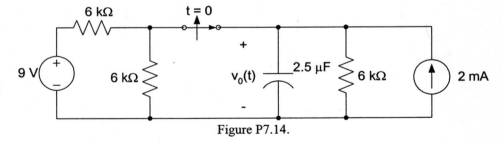

Figure P7.14.

121

P7.20. Use the differential equation approach to find i(t) for t > 0.

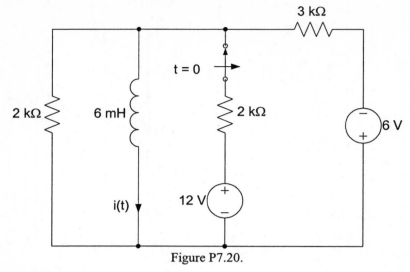

Figure P7.20.

P7.34. Find $v_0(t)$ for t > 0 using the step-by-step method.

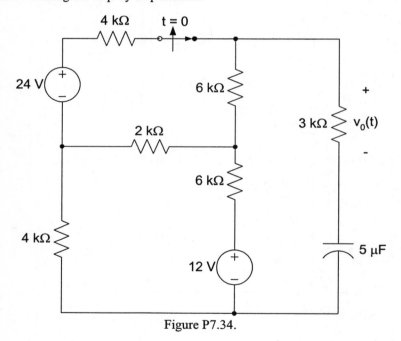

Figure P7.34.

P7.43. Find $i_0(t)$ for $t > 0$ using the step-by-step method.

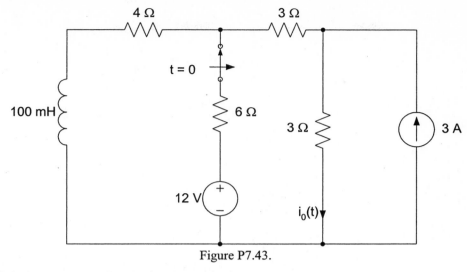

Figure P7.43.

P7.55. Find $i_0(t)$ for $t > 0$ using the step-by-step method.

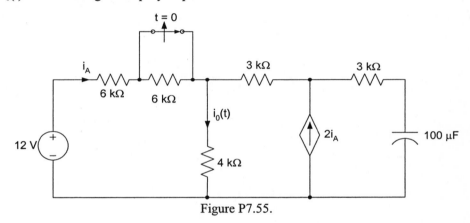

Figure P7.55.

P7.85. Find $v_C(t)$ for $t > 0$.

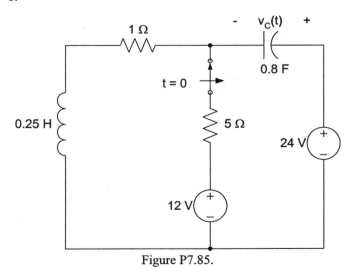

Figure P7.85.

P7.89. Find i(t) for t > 0.

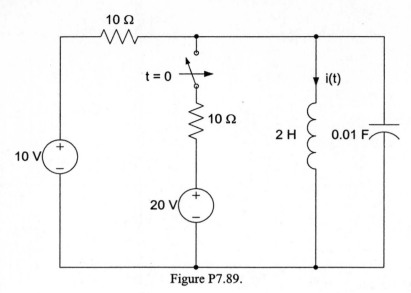

Figure P7.89.

CHAPTER 8

P8.13. Find **Z**.

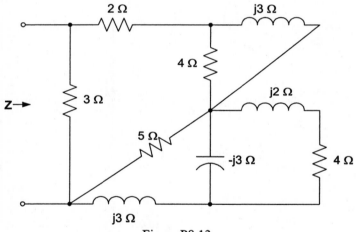

Figure P8.13.

P8.23. Find the value of C such that v(t) and i(t) are in phase.

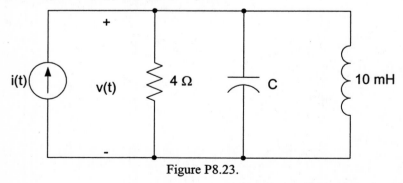

Figure P8.23.

P8.40. Find V_0.

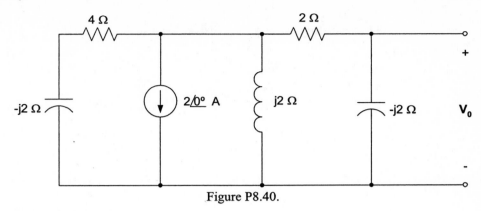

Figure P8.40.

P8.53. Find I_1 using nodal analysis.

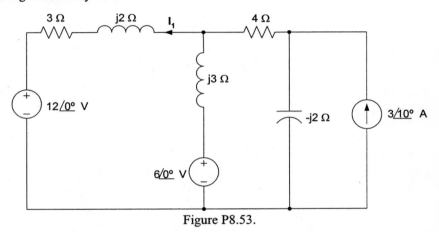

Figure P8.53.

P8.60.(a) Find V_x using nodal analysis.

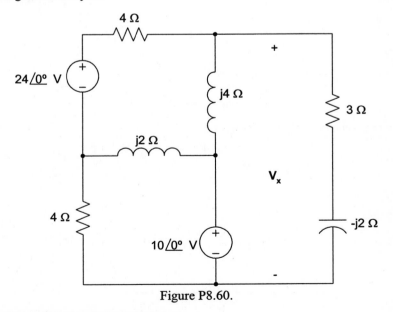

Figure P8.60.

P8.60.(b) Find V_x in Figure P8.60 using mesh analysis.

P8.69. Find V_0 using mesh analysis.

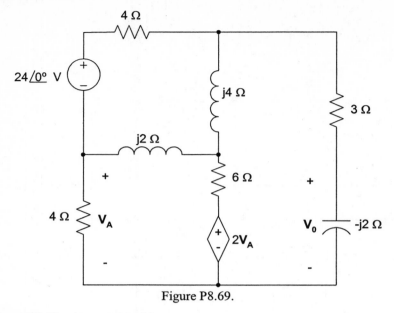

Figure P8.69.

P8.75. Find I_1 in Figure P8.53 using superposition.

P8.94. Find V_0 in Figure P8.69 using Thevenin's Theorem.

CHAPTER 9

P9.10. Find the power supplied and the average power absorbed by each element.

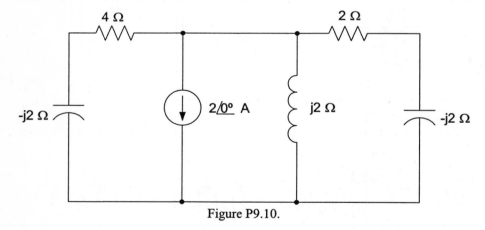

Figure P9.10.

P9.19. Determine the average power absorbed by the 4-Ω and 3-Ω resistors.

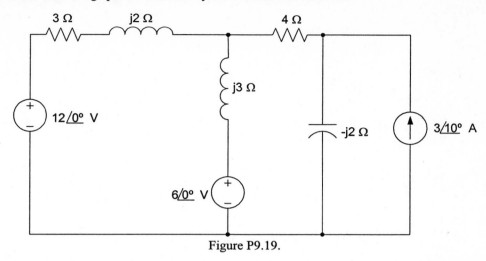

Figure P9.19.

P9.31. Determine Z_L for maximum average power transfer and the value of the maximum average power transferred to Z_L.

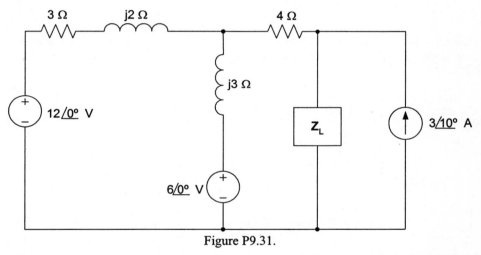

Figure P9.31.

P9.39. Find Z_L for maximum average power transfer and the value of the maximum average power transferred to Z_L.

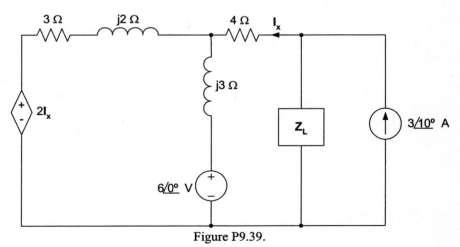

Figure P9.39.

P9.49. The voltage across a 2-Ω resistor is given by the waveform below. Find the average power absorbed by the resistor.

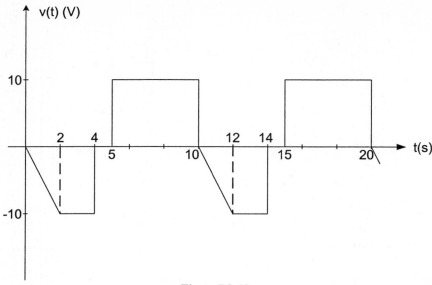

Figure P9.49.

P9.59. The source supplies 40 kW at a power factor of 0.9 lagging. The real and reactive losses of the transmission-line feeder are 1.6 kW and 2.1 kvar, respectively. Find the load voltage and the real and reactive power absorbed by the load.

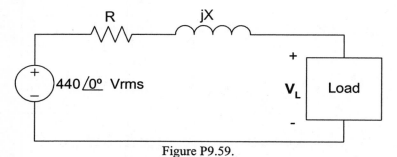

Figure P9.59.

P9.65. Find the power factor of the source and $v_S(t)$ if $f = 60$ Hz.

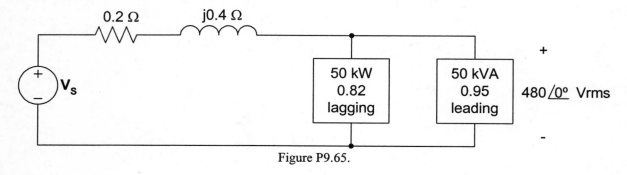

Figure P9.65.

128

P9.72. Find the value of capacitance to be connected in parallel with the load to make the source power factor to be 0.95 leading. f = 60 Hz.

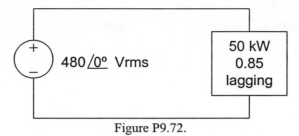

Figure P9.72.

CHAPTER 10

P10.7. Find V_0.

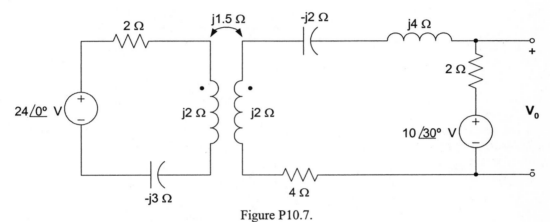

Figure P10.7.

P10.27. Find V_0.

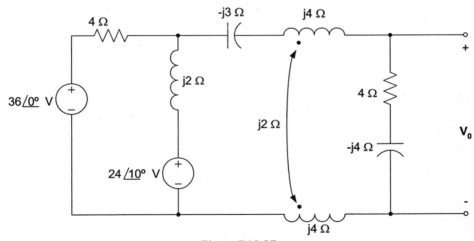

Figure P10.27.

P10.35. Find V_0.

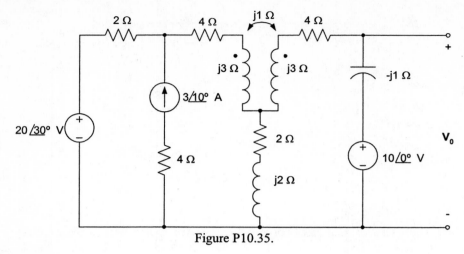

Figure P10.35.

P10.50. Determine V_0.

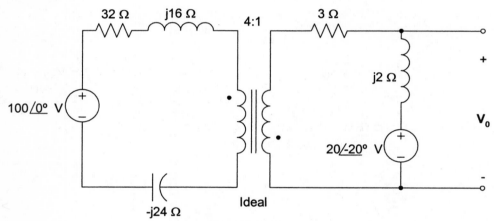

Figure P10.50.

P10.60. Determine V_0.

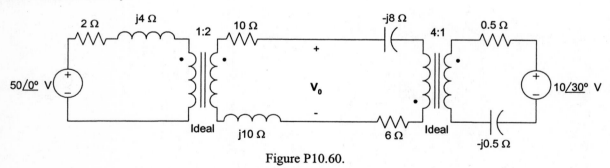

Figure P10.60.

P10.70. If $\mathbf{V_0} = 10\angle 30°$ V, find $\mathbf{V_S}$.

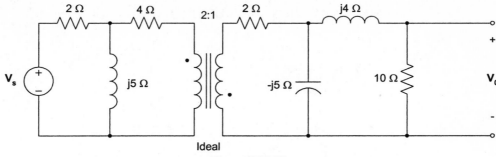

Figure P10.70.

CHAPTER 11

P11.6. A positive sequence balanced three-phase wye-connected source with a phase voltage of 277 Vrms supplies power to a balanced wye-connected load. The per-phase load impedance is $60 - j40\ \Omega$. Determine the line currents in the circuit if the phase angle of $\mathbf{V_{an}} = 0°$.

P11.10. An abc-sequence set of voltages feeds a balanced three-phase wye-wye system. The line and load impedances are $0.5 + j0.75\ \Omega$ and $20 - j24\ \Omega$, respectively. If the load voltage of the a-phase is $\mathbf{V_{AN}} = 125\angle 10°$ Vrms, find the line voltages of the input.

P11.19. In a balanced three-phase wye-wye system, the total power in the lines is 650 W. $\mathbf{V_{AN}} = 117\angle 15°$ Vrms and the power factor of the load is 0.88 leading. If the line impedance is $1 + j2\ \Omega$, determine the load impedance.

P11.29. Find the magnitude of the line voltage at the load.

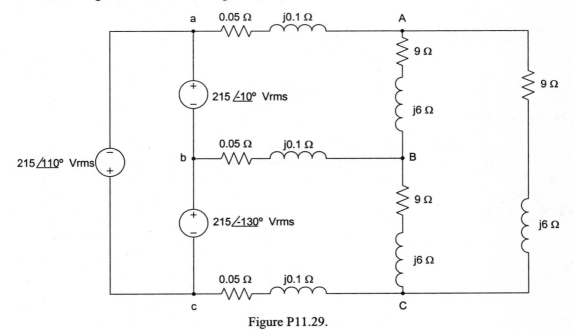

Figure P11.29.

P11.45. Find the line currents and the power absorbed by the delta-connected load.

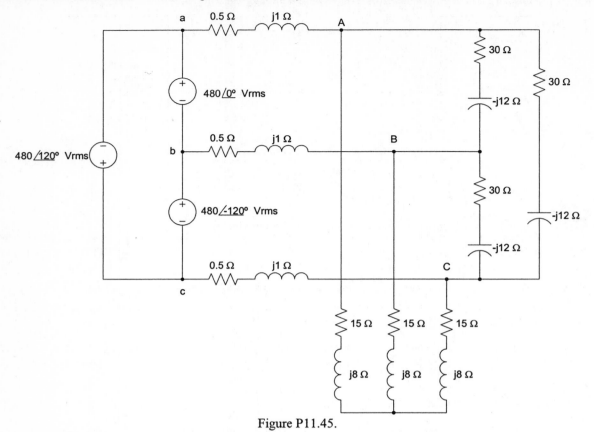

Figure P11.45.

P11.55. If the line voltage at the load is 480 Vrms, find the line voltage and power factor at the source.

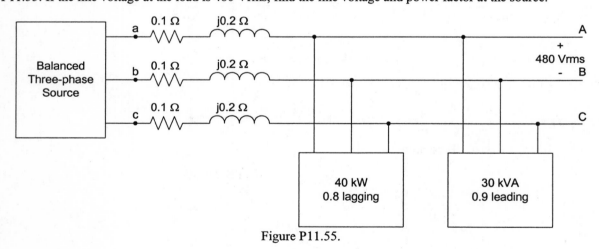

Figure P11.55.

P11.63. Find C such that the power factor of the source if 0.98 lagging.

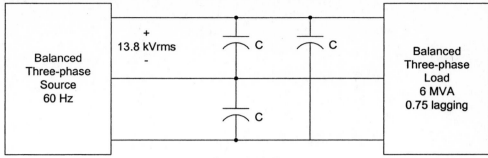

Figure P11.63.

CHAPTER 12

P12.2. Determine the voltage transfer function $V_o(s)/V_i(s)$ as a function of s.

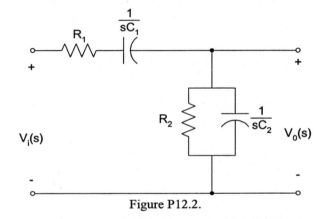

Figure P12.2.

P12.11 Sketch the magnitude characteristic of the Bode plot for the transfer function:

$$\mathbf{H}(j\omega) = \frac{5(j\omega + 10)}{j\omega(j\omega + 100)}.$$

P12.29. Find $\mathbf{H}(j\omega)$ if its magnitude characteristic is as shown below.

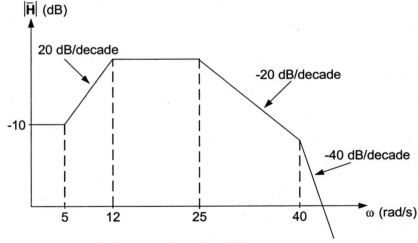

Figure P12.29.

P12.37. If the resonant frequency of the network is 10,000 rad/s, find L. Also compute the current at resonance, $\omega_0/3$, and $3\omega_0$.

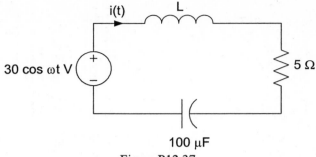
Figure P12.37.

P12.42. A parallel RLC resonant circuit has a resonant frequency of 12,000 rad/s and an admittance of 5 mS at resonance. Find R and C.

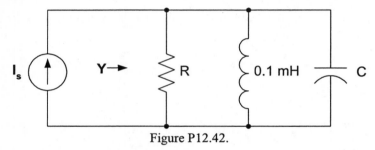
Figure P12.42.

P12.57. Determine what type of filter the network shown below represents by the determining the voltage transfer function.

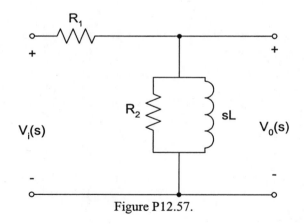
Figure P12.57.

CHAPTER 13

P13.7. Use the time shifting theorem to determine the Laplace transform of $f(t) = tu(t-1) + 3(t-2)u(t-2)$.

P13.11. Given $F(s) = \dfrac{s^2 + 5s + 1}{s(s+1)(s+4)}$, find f(t).

P13.21. Given $F(s) = \dfrac{4(s+3)}{(s+1)(s^2 + 2s + 5)}$, find f(t).

P13.23. Given $F(s) = \dfrac{100}{s^3(s+5)}$, find f(t).

P13.37. Use the Laplace transform to find y(t) if $\dfrac{dy}{dt} + 4y(t) + 4\int_0^t y(x)dx = 10u(t)$, $y(0) = 10$.

P13.42. Find the initial and final values of the time function f(t) if $\mathbf{F}(s) = \dfrac{8s^2 - 20s + 500}{s(s^2 + 4s + 50)}$.

P13.46. In the circuit below, the switch opens at t = 0. Use Laplace transforms to find $v_0(t)$ for t > 0.

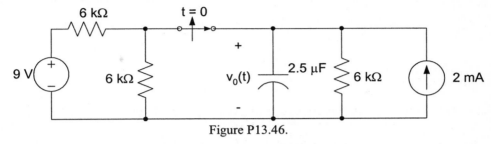

Figure P13.46.

P13.47. In the circuit below, the switch opens at t = 0. Use Laplace transforms to find i(t) for t > 0.

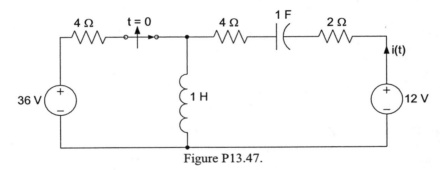

Figure P13.47.

CHAPTER 14

P14.6. Find $v_0(t)$ for t > 0 using nodal analysis.

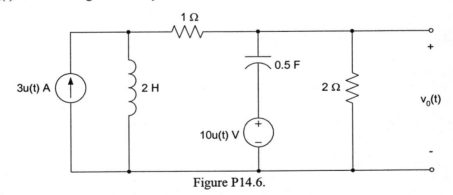

Figure P14.6.

P14.12. Find $v_0(t)$ for t > 0 in Figure P14.6 using mesh analysis.

P14.23. Use Thevenin's Theorem to determine $v_0(t)$ for t > 0 in Figure P14.6.

P14.25. Use Thevenin's Theorem to determine $v_0(t)$ for $t > 0$.

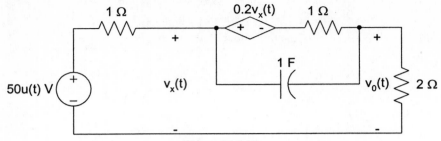

Figure P14.25.

P14.31. Find $i_0(t)$ for $t > 0$.

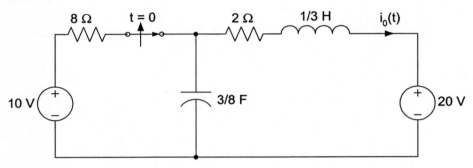

Figure P14.31.

P14.36. Find $v_0(t)$ for $t > 0$.

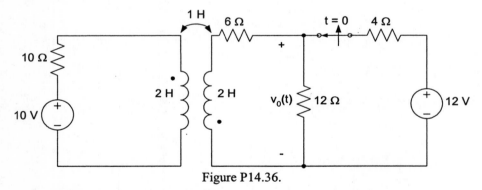

Figure P14.36.

P14.41. Determine the output voltage $v_0(t)$ if the input voltage $v_i(t)$ is given by the plot below.

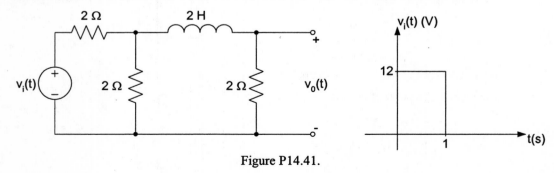

Figure P14.41.

P14.48. Find the transfer function for the network below.

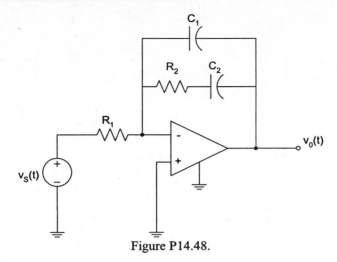

Figure P14.48.

P14.59. Find the steady-state response $v_{oss}(t)$.

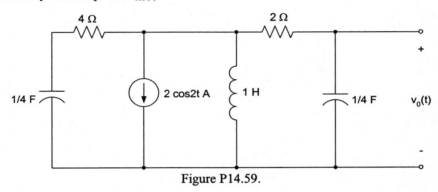

Figure P14.59.

CHAPTER 15

P15.2. Find the exponential Fourier series for the waveform shown below.

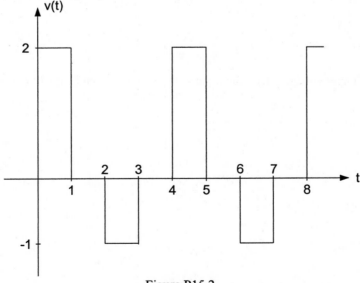

Figure P15.2.

P15.10. Determine the trigonometric Fourier series for the waveform shown in Figure P15.2.

P15.24. The discrete line spectrum for a periodic function is shown below. Determine the expression for f(t).

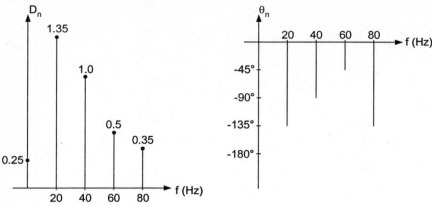

Figure P15.24.

P15.29. Determine the first three terms of the steady-state current i(t) if the input voltage is given by

$$v(t) = \frac{30}{\pi} + 15\sin 10t + \sum_{\substack{n=2 \\ n \text{ even}}}^{\infty} \frac{60}{\pi(1-n^2)} \cos 10nt \text{ V}.$$

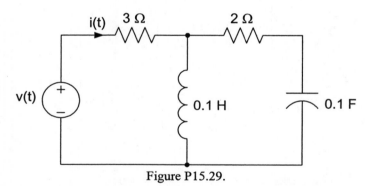

Figure P15.29.

P15.31. Find the average power absorbed by the network shown below if v(t) = 20 + 5 cos377t + 3.5 cos(754t - 20°) V and i(t) = 1.2 cos(377t-30°) + 0.8 cos(754t+45°) A.

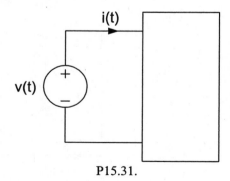

P15.31.

P15.37. Use the transform technique to find $v_0(t)$ in the network below if $v(t) = 15 \cos 10t$ V.

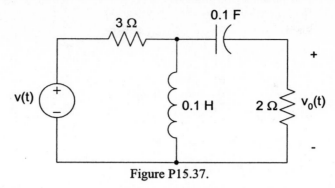

Figure P15.37.

P15.40. Determine the total 1-Ω energy content of the output $v_0(t)$ if $v_i(t) = 5e^{-2t}u(t)$ V.

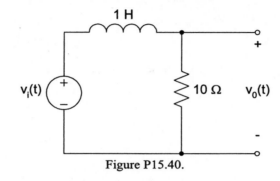

Figure P15.40.

CHAPTER 16

P16.3. Find the Y parameters for the two-port network shown below.

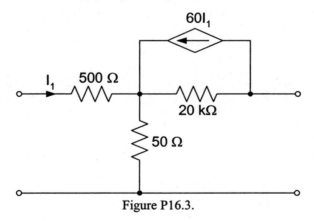

Figure P16.3.

P16.9. Determine the Z parameters for the two-port network shown below.

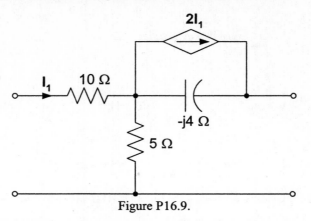

Figure P16.9.

P16.19. Find the hybrid parameters for the two-port network shown in Figure P16.3.

P16.24. Find the transmission parameters for the two-port network shown in Figure P16.9.

P16.42. Find $\mathbf{V_s}$ if $\mathbf{V_2} = 220\angle 0°$ Vrms in the network shown below.

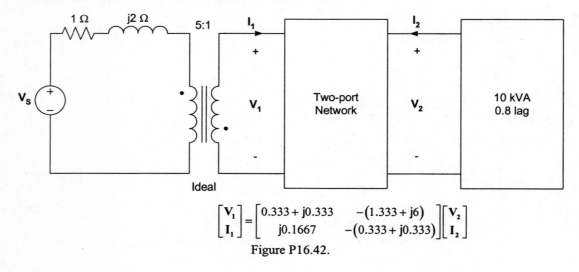

$$\begin{bmatrix} \mathbf{V_1} \\ \mathbf{I_1} \end{bmatrix} = \begin{bmatrix} 0.333 + j0.333 & -(1.333 + j6) \\ j0.1667 & -(0.333 + j0.333) \end{bmatrix} \begin{bmatrix} \mathbf{V_2} \\ \mathbf{I_2} \end{bmatrix}$$

Figure P16.42.

✍ Notes

✍ Notes

✍ Notes

✍ Notes

✍ Notes

✍ Notes

✍ Notes

✍ Notes

✍ Notes

✍ Notes

✍ Notes

✍ Notes

✍ Notes

✍ Notes

✍ **Notes**

✍ Notes

✍ Notes

✍ Notes

✍ Notes

✍ Notes

✍ Notes

✍ Notes